Improving Scientific Communication for Lifelong Learners

Gulsun Kurubacak-Meric
Anadolu University, Turkey

Serap Sisman-Ugur
Anadolu University, Turkey

A volume in the Advances in Higher Education and Professional Development (AHEPD) Book Series

IGI Global
PUBLISHER of TIMELY KNOWLEDGE

Published in the United States of America by
IGI Global
Information Science Reference (an imprint of IGI Global)
701 E. Chocolate Avenue
Hershey PA, USA 17033
Tel: 717-533-8845
Fax: 717-533-8661
E-mail: cust@igi-global.com
Web site: http://www.igi-global.com

Library of Congress Cataloging-in-Publication Data

Names: Kurubacak, Gulsun, 1964- editor. | Sisman-Ugur, Serap, 1981- editor.

Title: Improving scientific communication for lifelong learners / Gulsun Kurubacak-Meric and Serap Sisman-Ugur, editor.
Description: Hershey, PA : Information Science Reference, [2021] | Includes bibliographical references and index. | Summary: "Within the scientific community, much attention has focused on improving communications between scientists, policymakers, and the public. This book focuses on the efforts that are centered on improving the content, substantiality, accessibility, and delivery of scientific communications, and to convey clear information to an audience, so its members can understand, use and build on it"-- Provided by publisher.
Identifiers: LCCN 2020019404 (print) | LCCN 2020019405 (ebook) | ISBN 9781799845348 (hardcover) | ISBN 9781799858478 (paperback) | ISBN 9781799845355 (ebook)
Subjects: LCSH: Communication in science. | Scientific literature.
Classification: LCC Q223 .I44 2021 (print) | LCC Q223 (ebook) | DDC 501/.4--dc23
LC record available at https://lccn.loc.gov/2020019404
LC ebook record available at https://lccn.loc.gov/2020019405

This book is published in the IGI Global book series Advances in Higher Education and Professional Development (AHEPD) (ISSN: 2327-6983; eISSN: 2327-6991)

British Cataloguing in Publication Data
A Cataloguing in Publication record for this book is available from the British Library.

For electronic access to this publication, please contact: eresources@igi-global.com.

Advances in Higher Education and Professional Development (AHEPD) Book Series

ISSN:2327-6983
EISSN:2327-6991

Editor-in-Chief: Jared Keengwe University of North Dakota, USA

MISSION

As world economies continue to shift and change in response to global financial situations, job markets have begun to demand a more highly-skilled workforce. In many industries a college degree is the minimum requirement and further educational development is expected to advance. With these current trends in mind, the **Advances in Higher Education & Professional Development (AHEPD) Book Series** provides an outlet for researchers and academics to publish their research in these areas and to distribute these works to practitioners and other researchers.

AHEPD encompasses all research dealing with higher education pedagogy, development, and curriculum design, as well as all areas of professional development, regardless of focus.

COVERAGE

- Adult Education
- Assessment in Higher Education
- Career Training
- Coaching and Mentoring
- Continuing Professional Development
- Governance in Higher Education
- Higher Education Policy
- Pedagogy of Teaching Higher Education
- Vocational Education

IGI Global is currently accepting manuscripts for publication within this series. To submit a proposal for a volume in this series, please contact our Acquisition Editors at Acquisitions@igi-global.com or visit: http://www.igi-global.com/publish/.

The Advances in Higher Education and Professional Development (AHEPD) Book Series (ISSN 2327-6983) is published by IGI Global, 701 E. Chocolate Avenue, Hershey, PA 17033-1240, USA, www.igi-global.com. This series is composed of titles available for purchase individually; each title is edited to be contextually exclusive from any other title within the series. For pricing and ordering information please visit http://www.igi-global.com/book-series/advances-higher-education-professional-development/73681. Postmaster: Send all address changes to above address.

Titles in this Series

For a list of additional titles in this series, please visit:
http://www.igi-global.com/book-series/advances-higher-education-professional-development/73681

University-Industry Collaboration Strategies in the Digital Era
Durmuş Günay (Maltepe University, Turkey) Tuncer Asunakutlu (Izmir Bakircay University, Turkey) and Orkun Yildiz (Izmir Democracy University, Turkey)
Information Science Reference • © 2021 • 431pp • H/C (ISBN: 9781799839019) • US $195.00

Cultivating Entrepreneurial Changemakers Through Digital Media Education
Leslie Simone Byrd (Alabama State University, USA)
Information Science Reference • © 2021 • 273pp • H/C (ISBN: 9781799858089) • US $195.00

New Models for Technical and Vocational Education and Training
Moses Makgato (Tshwane University of Technology, South Africa) and George Afeti (Skills Initiative for Africa, Ghana)
Information Science Reference • © 2021 • 305pp • H/C (ISBN: 9781799826071) • US $175.00

Overcoming Fieldwork Challenges in Social Science and Higher Education Research
Abir El Shaban (Innovate Your Classroom (IC) Consulting, Canada)
Information Science Reference • © 2021 • 306pp • H/C (ISBN: 9781799858263) • US $195.00

Transforming Curriculum Through Teacher-Learner Partnerships
Pradeep Nair (Taylor's University, Malaysia) Michael James Keppell (Taylor's University, Malaysia) Chee Leong Lim (Taylor's University, Malaysia) TamilSalvi Mari (Taylor's University, Malaysia) and Nurhanim Hassan (Taylor's University, Malaysia)
Information Science Reference • © 2021 • 378pp • H/C (ISBN: 9781799864455) • US $195.00

For an entire list of titles in this series, please visit:
http://www.igi-global.com/book-series/advances-higher-education-professional-development/73681

IGI Global
PUBLISHER of TIMELY KNOWLEDGE

701 East Chocolate Avenue, Hershey, PA 17033, USA
Tel: 717-533-8845 x100 • Fax: 717-533-8661
E-Mail: cust@igi-global.com • www.igi-global.com

Yesterday I was smart,
I wanted to change the world;
Today I am wise,
I try to change myself.

-Rumi

Editorial Advisory Board

List of Reviewers

Table of Contents

Foreword xvii

Preface xix

Section 1
The Approaches

Chapter 1
Risk-Intelligent Communities for Lifelong Learners in a Post-Truth Era 1
Emre Dinçer, Adnan Menderes University, Turkey
Seval Kardeş Selimoğlu, Anadolu University, Turkey
Gülsün Kurubacak, Anadolu University, Turkey

Chapter 2
Design Thinking for Online Learning: A Design Thinking Approach 14
Soner Sözler, Zonguldak Bülent Ecevit University, Turkey

Chapter 3
Problem-Based Learning (PBL): A Deep Approach to Learning in the 21st Century 38
Mahmoud Hawamdeh, Al-Qud Open University, Palestine
Idris Adamu, Aminu Saleh College of Education, Nigeria

Chapter 4
Contributions of Information Design to Popular Scientific Communication: A Semiotic Approach 57
Juliana Rocha Franco, State University of Minas Gerais (UEMG), Brazil
Izabela Silva Pinho, State University of Minas Gerais (UEMG), Brazil

Section 2
Communities of Learning

Chapter 5
Presentation of a Social Responsibility Project on Social Media and MOOCs Within the Scope of Scientific Communication: Basic Health Knowledge, First Aid, and Coronavirus Example....................76
Gökhan Kuş, Anadolu University, Turkey

Chapter 6
Measuring Student Satisfaction Level Regarding Aesthetic and Instructor Dimensions of Web-Based Distance Learning Programs: The Case of Spiritual Guidance Training Program....................95
Serhat Koca, Anadolu University, Turkey
Murat Aytaş, Selcuk University, Turkey

Chapter 7
New Features of Learners in Education: Digital Awareness, Digital Competence, and Digital Fluency....................113
Figen Kilic, Mersin University, Turkey
Ismail Karakuş, Mersin University, Turkey

Chapter 8
Content-Generating Learners' Access to Accurate and Up-to-Date Information....................133
Hakan Kilinc, Anadolu University, Turkey
T. Volkan Yuzer, Anadolu University, Turkey

Chapter 9
Design of Communication Environments in Career-Focused Learning Communities....................156
Buket Kip Kayabaş, Anadolu University, Turkey

Chapter 10
Research and Network Opportunities in the Field of Science Communication....................180
Rabia Tas, Anadolu University, Turkey

Chapter 11
Scientific Research Articles: Twenty-Two Language Errors to Avoid....................195
John Blake, University of Aizu, Japan

Chapter 12
Reading the Narrative Structures in Educational Videos and Science Communication for the New Learner in the Age of Transhumanism................220
Basak Kalkan, Eskisehir Technical University, Turkey

Chapter 13
The Use of Animation and Storytelling in Science Communication.................240
Cigdem Tas Alicenap, Anadolu University, Turkey

Compilation of References 251

About the Contributors 282

Index 287

Detailed Table of Contents

Foreword xvii

Preface xix

Section 1
The Approaches

Chapter 1
Risk-Intelligent Communities for Lifelong Learners in a Post-Truth Era 1
Emre Dinçer, Adnan Menderes University, Turkey
Seval Kardeş Selimoğlu, Anadolu University, Turkey
Gülsün Kurubacak, Anadolu University, Turkey

After Oxford Dictionaries had announced the word "post-truth" as the Word of the Year in 2016, the concept of post-truth became highly popular in contemporary literacy. Use of the word became more widespread. Post-truth can be simply explained as disappearance of norms for truth. In a society divided by bias, feelings and personal believes are more important than objective facts. In the post truth era, realities don't make a difference. It is a challenging environment for public in this COVID-19 times and within traditional governance structures established around multiple layers of competence where lines of accountability and responsibility are not clear. For the future to come any time soon, we need risk intelligent communities that consist of lifelong learners.

Chapter 2
Design Thinking for Online Learning: A Design Thinking Approach 14
Soner Sözler, Zonguldak Bülent Ecevit University, Turkey

With the spread of COVID-19, nearly all education systems in the world pivoted to distance education in order to overcome the crisis. Because it is an unexpected crisis, most institutions had to employ emergency remote teaching systems. This decision brought so many problems to education industry. But now, it is a known fact that the virus is with us at least one more year, and moreover, new epidemics may occur. Therefore, this chapter aims to provide a guideline with a different approach to online learning for practitioners working in education industry. Thanks to this guideline, the practitioners can create well-designed online learning environments, and as a result, they will not need emergency remote teaching applications anymore.

Chapter 3
Problem-Based Learning (PBL): A Deep Approach to Learning in the 21st Century....................38
Mahmoud Hawamdeh, Al-Qud Open University, Palestine
Idris Adamu, Aminu Saleh College of Education, Nigeria

This chapter discuss how Problem-Based learning (PBL) helps to achieve this century's approach to teaching and learning for students in higher educational institutions. If adopted, this method of teaching will enable student to attain learning skills (skills, abilities, problem solving, and learning dispositions that have been identified) to acquire a lifelong habit of approaching problems with initiative and diligence and a drive to acquire the knowledge and skills needed for an effective resolution. And they will develop a systematic approach to solving real-life problems using higher-order skills.

Chapter 4
Contributions of Information Design to Popular Scientific Communication: A Semiotic Approach....................57
Juliana Rocha Franco, State University of Minas Gerais (UEMG), Brazil
Izabela Silva Pinho, State University of Minas Gerais (UEMG), Brazil

In a world increasingly saturated with data, scientific communication's success in reaching its target audience is linked to its ability to organize and represent information. Through semiotics concepts, this chapter explores the possibilities of information design as a set of tools for planning the exhibition of scientific content effectively. Information design is an area of knowledge that focuses on the human component's concern and the ergonomic aspects of visual communication. The chapter brings two case studies on infographics and on how information design is applied in popular scientific communication. The research will show that information design, associated with the precepts of plain language, can offer substantial contributions to the work of those who intend to disseminate science to non-specialized audiences.

Section 2
Communities of Learning

Chapter 5
Presentation of a Social Responsibility Project on Social Media and MOOCs Within the Scope of Scientific Communication: Basic Health Knowledge, First Aid, and Coronavirus Example..76
Gökhan Kuş, Anadolu University, Turkey

The learning activities of individuals are not only limited to school, but they also continue learning in all areas of life. One of the most effective ways of achieving this in the evolving socio-cultural and technological environment is the social media environment. The authors addressed the issues separately for the deficiencies identified in the study. They offered the trainings they have done in various platforms.

Chapter 6
Measuring Student Satisfaction Level Regarding Aesthetic and Instructor Dimensions of Web-Based Distance Learning Programs: The Case of Spiritual Guidance Training Program..95
Serhat Koca, Anadolu University, Turkey
Murat Aytaş, Selcuk University, Turkey

Today, many educational institutions provide their content to their recipients through web-based distance learning applications. While providing this education method to the recipients through communication technologies, the aesthetic design and the instructor dimension of the content are of utter importance. In this framework, the study measured the level of student satisfaction regarding the aesthetic and instructor dimension in web-based distance learning programs within the case of the Spiritual Guidance program. Accordingly, the study of distance learning programs in terms of aesthetic and instructive dimensions will contribute to the research that will be conducted in this direction.

Chapter 7
New Features of Learners in Education: Digital Awareness, Digital Competence, and Digital Fluency..113
Figen Kilic, Mersin University, Turkey
Ismail Karakuş, Mersin University, Turkey

Throughout the recent decades there have been rapid developments and changes in terms of science and technology. From now on, individuals want a flexible education format in which time and place are changeable in line with their desires and concerns. Particularly from the perspective of lifelong learning, digitalization can be seen as an important element which forces individuals to change. Some of these are digital awareness, digital competence, and digital fluency. As one of the digital terms, digital competence includes the effective, efficient, critical, flexible, and reflective structuring of information for business, learning, and socializing, while the latter digital fluency is the ability to know when and where to use technology. The main problem that stands out is adapting education programs according to the new developments. The restructuring of the program development activities, so as to raise individuals with changing and developing knowledge, skills, and attitudes, acts like a spotlight which highlights the critical role of program development in the education system.

Chapter 8
Content-Generating Learners' Access to Accurate and Up-to-Date Information 133
Hakan Kilinc, Anadolu University, Turkey
T. Volkan Yuzer, Anadolu University, Turkey

This study aims to reveal the characters and responsibilities of content generating learners, as new generation learners, in the learning processes for producing the contents which contain accurate and up to date information. In this study, which was designed as a case study for this purpose, the opinions of the field experts who voluntarily participated in the study were obtained about what features the content generating learners should have at the point of accessing accurate and up-to-date information. The answers given by the field experts to the research questions directed to them were analyzed through the NVIVO 11 package program, and the themes reached were presented within the scope of the study. It is thought that this study will contribute to the related literature in order to ensure that the content generating learners can produce more accurate and more up-to-date content.

Chapter 9
Design of Communication Environments in Career-Focused Learning Communities.. 156
Buket Kip Kayabaş, Anadolu University, Turkey

Rapid renewal of information with the rapid development of technology prevents individuals from using their knowledge and skills throughout their lives. Today, instead of participating in long education programs such as undergraduate or associate degree programs, short-term education programs that are specialized according to their fields of study are increasingly in demand. Investigating which communication environments individuals prefer in career focused learning communities in which individuals participate in lifelong learning is important in terms of shedding light on the design process of these environments. The purpose of this manuscript is firstly to reveal the opportunities offered by lifelong learning to individuals, then to determine the preferences of the communication medium based on the experiences of individuals participating in career focused learning communities offered within the scope of lifelong learning, and finally, to provide suggestions for the design of communication environments to people who create and manage career focused learning communities.

Chapter 10
Research and Network Opportunities in the Field of Science Communication 180
Rabia Tas, Anadolu University, Turkey

Science and technology have been developing at a great pace. There are numerous breakthrough innovations and groundbreaking improvements in almost every aspect of life. Undoubtedly, another field experiencing such a fast development is communication. Communication has been following and adapting related technological developments not only in terms of software and hardware but also in terms of targeting the right audience and creating the audience interaction. While developments in science and technology feed communication, communication disseminates the outcomes of innovative studies to the defined target groups. Science communication becomes a part of this mutual relation. This study is focused on research funds, opportunities, and international networks in the field of science communication. Within this context, challenges in research activities and expected impact of research outcomes will be examined in terms of science communication. Also, the role and importance of networks in science communication will be examined in line with research funds and support provided for this field.

Chapter 11
Scientific Research Articles: Twenty-Two Language Errors to Avoid............... 195
John Blake, University of Aizu, Japan

Error-free scientific research articles are more likely to be accepted for publication than those permeated with errors. This chapter identifies, describes, and explains how to avoid 22 common language errors. Scientists need to master the genre of scientific writing to conform to the generic expectations of the community of practice. Based on a systematic analysis of the pedagogic literature, five categories of errors were identified in scientific research articles namely accuracy, brevity, clarity, objectivity, and formality. To gain a more in-depth understanding of the errors, a corpus investigation of scientific articles was conducted. A corpus of 200 draft research articles submitted for internal review at a research institute with university status was compiled, annotated, and analyzed. This investigation showed empirically the types of errors within these categories that may impinge on publication success. In total, 22 specific types of language errors were identified. These errors are explained, and ways for overcoming each of them are described.

Chapter 12
Reading the Narrative Structures in Educational Videos and Science Communication for the New Learner in the Age of Transhumanism................220
Basak Kalkan, Eskisehir Technical University, Turkey

When defining the new world, where knowledge is not independent of the one who possesses knowledge, it is considered imperative to develop new models for learning and science communication processes that are not simply confined to dichotomies such as tangible and intangible, or measurable and immeasurable. In the age of transhumanism, the concepts of production and consumption are freed from the shackles of absolute conceptualization and are defined instead with the concepts of producing consumer or consuming producer. In the science communication process, the new learner must be defined without being confined to absolute definitions on the new road map that is defined for the subject. The new learner is the actor, observer, adventurer, creative nomadic individual who cannot be trapped with absolutes in the post-modern era. This study seeks an answer to the question of whether or not a new content design is possible for the actor, observer, adventurer, creative nomadic individual in the face of the formlessness, ambiguity, and continuity of postmodernism.

Chapter 13
The Use of Animation and Storytelling in Science Communication.................240
Cigdem Tas Alicenap, Anadolu University, Turkey

Science communication has brought new teaching/learning environments forward. Various digital media have provided appropriate platforms. Learning processes for everyone have been differentiated and thrived on newly developed technologies and digital media. Animation is among these to provide an alternative to science communication methods. Especially animations developed with storytelling can become very effective in learning processes. In cartoon movies, certain knowledge on different fields has been explained by turning them into scenarios. It's possible to see examples of cartoons in this concept in Turkey and all around the world. Storytelling (scenario) and animation dynamics make knowledge more impressive and long-lasting. Educational cartoon movies produced for children, which aren't in formal and organized education but younger than the school age, have been produced in a widespread manner. This study is focused on transferring scientific knowledge by using animation products. In this context, educational cartoon movies will be examined through storytelling and animation.

Compilation of References 251

About the Contributors 282

Index 287

Foreword

Scientific communication is one of the strong pillars of scientific research. Effective communication is a hallmark for scientific writing. It helps us in presenting our complex ideas in simple language with clarity and concision. The cycle of scientific communication consists of various stages, viz, production, transmission and advancement. Through thinking, experimenting, and researching, we produce knowledge, which then is passed on from generation to generation. As famously said by Sir Isaac Newton, "If I have seen further, it is by standing on the shoulders of giants". Since then, and even before that, the academic and scientific writing has metamorphosed greatly. From hand copying the text, to development of printing press and then arrival of computers and Internet has transformed the field all together.

The third decade of the 21st Century has witnessed two major revolutions impacting education and learning: the 4th Industrial Revolution and the 4th Education Revolution (highlighted by Sir Anthony Seldon in his book by the same name). These two revolutions have necessitated an important role for all stakeholders in any educational system where we need to be a lifelong learner. For a lifelong learner, the static learning has to change to dynamic learning. Around two decades ago, David Nunan proposed five levels of fostering learner autonomy: awareness, intervention, involvement, creation and transcendence. This calls for such systems which support autonomous learner. Scientific communication is the key to autonomous learning systems.

Technological developments like Artificial Intelligence, Swarm Intelligence, Machine Learning and Quantum Technologies are already transforming education. Self-learning is one of the most crucial skills for being future ready. As the knowledge is going inter-disciplinary and cross-disciplinary, it is important to ponder over how to organise knowledge. Knowledge representation and knowledge organisation become a determining force for scientific writings and communication. A successful scientific communication depends upon how the message is delivered. It must clearly convey what the author wants to pass on to the readers. This brings in certain skill: the writing skills, communication skill, argumentation, critiques, composition, and expression. These call upon a great focus on Precision, Simplicity, Concision and

Fluidity. Any great research will not be impactful, unless it is published, which in turn depends upon good scientific writing. Effective knowledge organisation and in appropriate language brings its rewards.

The 4th Industrial Revolution technologies have resulted in big changes during the past few decades the way work and live. What is needed all the more importantly is the change in the way we write scientific papers. Scientific communication has been around for more than 300 years, but the pace in the number of scientific outputs increased post World War II. This has further been fuelled by introduction of digital technologies. These digital technologies changed the medium of content, nature of content and the outreach. Another development noted has been the rise of new publishing enterprises. With increasing number of publishable researches, such publishing enterprises enforced norms for quality publishing and standards. Roles and responsibilities of authors, editors and reviewers have been refined. Rigorous processes of quality check and ethical code of conduct for the scientific writing and publishable work have been put in place. A proper dissemination of a scientific work is an integral part of the research process, in the form of well communicated publication. The phrase 'Publish or Perish' coined by Coolidge in 1932 is a dilemma for every academic. It has long been an obligation for academics, for performance appraisals, and new career opportunities. Publishing is an important step for advancement of knowledge, what makes it crucial is to get published in good publications. The number of papers the journal and book editors receive is much higher and thus poorly written ones are filtered out by them in the initial stage. Those who pass this stage are then subjected to blind or peer review process to sort out the best among the equals. This is certainly a challenge for everyone: authors, editors, and reviewers.

Here's the book to which covers the ABC of scientific communication (accuracy, brevity and clarity). The book covers interesting, timely and relevant topics ranging from applications to distance education, design thinking for online learning to applications of artificial intelligence. It covers storytelling, animation, digital awareness, digital competence, digital fluency, student satisfaction in web-based distance learning programs, social responsibility for basic health knowledge, educational videos, learning communities, visual communication, problem-based learning, and how to produce error-free scientific research articles.

The editors have done an excellent job in this volume, I congratulate them and the authors of various chapters. This book would be a great treasure for those who wish to learn the art and science of scientific communication.

Ramesh C. Sharma
Ambedkar University, Delhi, India

Preface

Scientific Communication (Sci-Com) is a part of information science and the sociology of science that study researchers' use of formal and informal information channels as well as their communicative roles. the utilization of the formal publication system and similar issues. Therefore, the main purpose of the edited Sci-Com book is to improve scientific writing and speaking skills for lifelong learning researchers by developing an adaptive and responsive open and distance application according to universal design principles.

This book, consisting of 13 chapters, is divided into two sections: "The Approaches" and "Communities and Learning."

SECTION 1: THE APPROACHES

Chapter 1: Risk Intelligent Communities for Lifelong Learners in Post-Truth Era

After Oxford Dictionaries had announced the word "post-truth" as the Word of the Year in 2016, concept of post-truth became highly popular in contemporary literacy. Use of the word became more widespread. Post-truth can be simply explained as disappearance of norms for truth. In a society divided by bias, feelings and personal believes are more important than objective facts. In the post truth era, realities don't make a difference. It is a challenging environment for public in this Covid-19 times. Within traditional governance structures established around multiple layers of competence where lines of accountability and responsibility are not clear. For the future to come any time soon, we need risk intelligent communities that consist of lifelong learners.

Chapter 2: Design Thinking for Online Learning – A Design Thinking Approach

With the spread of Covid-19 nearly all education systems in the world pivoted to Distance Education in order to overcome the crisis. Because it is an unexpected crisis, most institutions had to employ emergency remote teaching systems. This decision brought so many problems to education industry. But now, it is a known fact that, the virus is with us at least one more year and moreover new epidemics may occur. Therefore, this chapter aims to provide a guideline with a different approach to Online Learning for practitioners working in education industry. Thanks to this guideline, the practitioners can create well-designed Online Learning environments and as a result they will not need emergency remote teaching applications anymore.

Chapter 3: Problem-Based Learning (PBL) – A Deep Approach to Learning in 21st Century

This chapter discuss how PBL help to achieve this century approach to teaching and learning for students in higher educational institutions if adopt this method of teaching through their teachers in order to attain learning skills (skills, abilities, problem solving and learning dispositions that have been identified) to acquire a lifelong habit of approaching problems with initiative and diligence and a drive to acquire the knowledge and skills needed for an effective resolution and develop a systematic approach to solving real-life problems using higher-order skills.

Chapter 4: Contributions of Information Design to Popular Scientific Communication – A Semiotic Approach

In a world increasingly saturated with data, the success of scientific communication in reaching its target audience is linked to its ability to organize and represent information. This work explores, through concepts from Semiotics, the possibilities of Information Design as a set of tools for planning the exhibition of scientific content effectively. Information Design is an area of knowledge that focuses on the concern with the human component, and ergonomic aspects of visual communication. The research will show that Information Design associated with the precepts of clear language, can offer substantial contributions to the work of those who intend to disseminate science to non-specialized audiences. The chapter brings two case studies on the use of infographics (one of how Information Design is applied) in scientific dissemination, more precisely in the field of Entomology.

SECTION 2: COMMUNITIES AND LEARNING

Chapter 5: Presentation of a Social Responsibility Project on Social Media and MOOCs Within the Scope of Scientific Communication Basic Health Knowledge, First Aid, and Coronavirus Example

The learning activities of individuals are not only limited to school, but they also continue learning in all areas of life. We believe to be deficiencies in these areas and we will handle the subjects of "Basic Health Knowledge and First Aid" and "coronavirus, which has become a current issue as of the end of 2019 and the beginning of 2020". A new health problem that has affected the whole world has appeared as of late 2019 and early 2020: a corona virus known as COVID19.One of the most effective ways of achieving this in the evolving socio-cultural and technological environment is the social media environment. We addressed the issues separately for these deficiencies identified in our study. We offered the trainings we have done in various platforms; "basic health and first aid in "Explain in 90 Seconds Project" in Social Media; "Basic First Aid Knowledge and Coronavirus in Akadema".

Chapter 6: Measuring Student Satisfaction Level Regarding Aesthetic and Instructor Dimensions of Web-Based Distance Learning Programs – The Case of Spiritual Guidance Training Program

Today, many educational institutions provide their content to their recipients through web-based distance learning applications. While providing this education method to the recipients through communication technologies, the aesthetic design, and the instructor dimension of the content is of utter importance. In this framework, the study measured the level of student satisfaction regarding the aesthetic and instructor dimension in web-based distance learning programs within the case of the Spiritual Guidance program. Accordingly, the study of distance learning programs in terms of aesthetic and instructive dimensions will contribute to the research that will be conducted in this direction.

Chapter 7: New Features of Learners in Education Digital Awareness, Digital Competence, and Digital Fluency

Throughout the recent decades there have been rapid developments and changes in terms of science and technology. From now on, individuals want a flexible education format in which time and place are changeable in line with their desire and concerns. Particularly from the perspective of lifelong learning, digitalization can be seen as an important element which forces individuals to change. Some of these are digital awareness, digital competence and digital fluency. As one of the digital terms digital competence includes the effective, efficient, critical, flexible and reflective structuring of information for business, learning and socializing, while the latter digital fluency, is the ability to know when and where to use technology. The main problem that stands out is adapting education programs according to the new developments. The restructuring of the program development activities, so as to raise individuals with changing and developing knowledge, skills and attitudes, acts like a spotlight which highlights the critical role of program development in the education system.

Chapter 8: Content Generating Learners' Access to Accurate and Up-to-Date Information

This study aims to reveal the characters and responsibilities of content generating learners', as new generation learners, in the learning processes for producing the contents which contain accurate and up to date information. In this study, which was designed as a case study for this purpose, the opinions of the field experts who voluntarily participated in the study were obtained about what features the content generating learners should have at the point of accessing accurate and up-to-date information. The answers given by the field experts to the research questions directed to them were analyzed through the NVIVO 11 package program and the themes reached were presented within the scope of the study. It is thought that this study will contribute to the related literature, in order to ensure that the content generating learners can produce more accurate and more up-to-date content.

Chapter 9: Design of Communication Environments in Career-Focused Learning Communities

Rapid renewal of information with the rapid development of technology prevents individuals from using their knowledge and skills throughout their lives. Today, instead of participating in long education programs such as undergraduate or associate degree programs, short-term education programs that are specialized according to their fields of study are increasingly in demand. Investigating which communication environments individuals prefer in career focused learning communities in which individuals participate in lifelong learning is important in terms of shedding light on the design process of these environments. The purpose of this manuscript is firstly to reveal the opportunities offered by lifelong learning to individuals; then determine the preferences of the communication medium based on the experiences of individuals participating in career focused learning communities offered within the scope of lifelong learning; and finally, to provide suggestions for the design of communication environments to people who create and manage career focused learning communities.

Chapter 10: Research and Network Opportunities in the Field of Science Communication

Science and technology have been developing at a great pace. There are numerous breakthrough innovations and groundbreaking improvements in almost every aspect of life. Undoubtedly, another field experiencing such a fast development is communication. Communication has been following and adapting related technological developments not only in terms of software and hardware but also in terms of targeting the right audience and creating the audience interaction. While developments in science and technology feed communication, communication disseminates the outcomes of innovative studies to the defined target groups. Science communication becomes a part of this mutual relation. This study is focused on research funds, opportunities and international networks in the field of science communication. Within this context, challenges in research activities and expected impact of research outcomes will be examined in terms of science communication. Also, the role and importance of networks in science communication will be examined in line with research funds and support provided for this field.

Chapter 11: Scientific Research Articles – Twenty-Two Language Errors to Avoid

Error-free scientific research articles are more likely to be accepted for publication than those permeated with errors. This chapter identifies, describes and explains how to avoid twenty-two common language errors. Scientists need to master the genre of scientific writing to conform to the generic expectations of the community of practice. Based on a systematic analysis of the pedagogic literature, five categories of errors were identified in scientific research articles namely accuracy, brevity, clarity, objectivity and formality. To gain a more in-depth understanding of the errors, a corpus investigation of scientific articles was conducted. A corpus of two-hundred draft research articles submitted for internal review at a research institute with university status was compiled, annotated and analyzed. This investigation showed empirically the types of errors within these categories that may impinge on publication success. In total, twenty-two specific types of language errors were identified. These errors are explained and ways for overcoming each of them are described.

Chapter 12: Reading the Narrative Structures in Educational Videos and Science Communication for the New Learner in the Age of Transhumanism

When defining the new world, where knowledge is not independent of the one who possesses knowledge, it is considered imperative to develop new models for learning and science communication processes that are not simply confined to dichotomies such as tangible and intangible, or measurable and immeasurable. In the age of transhumanism, the concepts of production and consumption are freed from the shackles of absolute conceptualization, and are defined instead with the concepts of producing consumer or consuming producer. In the science communication process, the new learner must be defined without being confined to absolute definitions on the new road map that is defined for the subject. The new learner is the actor, observer, adventurer, creative nomadic individual who cannot be trapped with absolutes in the post-modern era.. This study seeks an answer to the question whether or not a new content design is possible for the actor, observer, adventurer, creative nomadic individual in the face of the formlessness, ambiguity and continuity of postmodernism.

Chapter 13: The Use of Animation and Storytelling in Science Communication

Science communication has brought new teaching/learning environments forward. Various digital media have provided appropriate platforms. Learning processes for everyone have been differentiated and thrived on newly developed technologies and digital media. Animation is among these to provide an alternative to science communication methods. Especially animations developed with storytelling can become very effective in learning processes. In cartoon movies, certain knowledge on different fields has been explained by turning them into scenarios. It's possible to see examples of cartoons in this concept in Turkey and all around the world. Storytelling (scenario) and animation dynamics make knowledge more impressive and long-lasting. Educational cartoon movies produced for children, which aren't in formal and organized education but younger than the school age, have been produced in a widespread manner. This study is focused on transferring scientific knowledge by using animation products. In this context, educational cartoon movies will be examined through storytelling and animation.

Learning can no longer be divided into place and time of school and workplace. Today's people are be exposed to with more information than they can handle. In the near future, people will need to know much more than now. Lifelong learning is a fundamental challenge and necessity to building the future of our societies. Lifelong learning is more than adult education and or just education - it is a mindset and a habit which people must adopt. This publication can be used for researchers, mentors, facilitators and tutors as well as learners. As it covers the management, communication, pedagogy, technology/computing, evaluation, biology/genetics/psychology, and sociology/intelligent social networks-based future trends, issues and challenges of learning in the age of transhumanism, the chapters answer their questions in this context. Because of lifelong learning as a design for the future, this publication is a course supplement book in not only sociology, computer science, psychology, etc. but also open education, open and distance education, online education, online learning, transformative online learning, and transformative learning. In addition, it is a unique resource for the researchers, scholars and professionals and lifelong learners. This proposed book is a reference book, and also a welcome addition to academic libraries' collections in natural and life science as well as social science.

Section 1
The Approaches

Chapter 1
Risk-Intelligent Communities for Lifelong Learners in a Post-Truth Era

Emre Dinçer
Adnan Menderes University, Turkey

Seval Kardeş Selimoğlu
Anadolu University, Turkey

Gülsün Kurubacak
Anadolu University, Turkey

ABSTRACT

After Oxford Dictionaries had announced the word "post-truth" as the Word of the Year in 2016, the concept of post-truth became highly popular in contemporary literacy. Use of the word became more widespread. Post-truth can be simply explained as disappearance of norms for truth. In a society divided by bias, feelings and personal believes are more important than objective facts. In the post truth era, realities don't make a difference. It is a challenging environment for public in this COVID-19 times and within traditional governance structures established around multiple layers of competence where lines of accountability and responsibility are not clear. For the future to come any time soon, we need risk intelligent communities that consist of lifelong learners.

DOI: 10.4018/978-1-7998-4534-8.ch001

INTRODUCTION

As Ohanian cites (2008), famous 20th century novelist George Orwell said that "In times of universal deceit, telling the truth will be a revolutionary act." It seems that Orwell was quite right about his prediction. After Oxford Dictionaries had announced the word "post-truth" as the Word of the Year in 2016, concept of post-truth became highly popular in contemporary literacy. Use of the word became more widespread. Post-truth can be simply explained as disappearance of norms for truth. In a society divided by bias, feelings and personal believes are more important than objective facts. In the post truth era, realities don't make a difference. Facts can be hidden or introduced in a political setting that bends the truth in its favour. One might be tempted to blame politicians for post-truth but what if the facts are far beyond politics.

HOW TO SURVIVE INTELLECTUALLY IN POST TRUTH ERA

One might have happened to notice that in social media platforms, it has become highly fashionable that people tell "white lies" about their accomplishments, places they travelled, their financial situation and of course their physical appearance. People prefer to exaggerate every positive aspect of their lives. Everyone wants to turn their dreams into realities but that is just not possible. Instead of facing the truth, we run into a dream world that can easily be accessed via technology. The technology has enabled human beings to go beyond their intellectual capacity. And humanity is under siege with information. There is so much information that attacks us that it feels like we are losing control of our lives.

John Ray (born, 1627 — died, 1705), leading 17th-century English naturalist and botanist who contributed fundamentally to advance in scientific categorization (Britannica, 2020) proposes that "Learning makes the wise wiser and the fool more foolish". Thinking about the chaotic times we live in, Ray's quote shed light on the paradox prevailing common sense. Let us explain that: Technological advancements, we own, haven't made us wiser. It is obvious that because of easy use of information technology, everyone seems to know everything but this hasn't wiped out the ignorance in society at all. Fake news and false information govern people with their fear. That explains why we need wisdom and wise people because without wisdom one can't find his way to enlightenment. A first step to fight ignorance might be to start learning from the wise. Learning from the wise can't be just put into practice with easy steps. It can only be done with a system that supports cultural products to make society interiorize lives of the wise. A government cultural policy with help of popular culture will be able to yield cultural products that favour the wisdom which might pave the way goes to enlightenment.

Intellectual poverty is like a son of laziness. If we don't use logic and develop critical thinking abilities in our daily life, we might end up in intellectual poverty. Act of survival itself includes lots of effort in it. That being said; if one longs to survive intellectually in post-truth, he or she must have the will to seek the truth. Every day, we encounter lots of news floating in social media environments. They are shared and spread like virus. If we can't make time for seeking the truth, how can we avoid the misleading information? What if we are manipulated? This would appear to indicate that a great deal of attention must be paid when we confront sceptical information.

Democracies have always trusted in common sense. In 1776, Thomas Paine stated that governments depend for its legitimacy entirely on the consent of the society. This idea was the ideological pillar that democracies stood on. Legitimate governments have always sought approval of the people that they intend to govern until now. In 2018 newspaper Times reported that Cambridge Analytica - British political consulting firm was eager to offer psychological profiles of American voters to political missions. This firm obtained the private Facebook data of tens of millions of users. After this scandal was revealed, it turned out to be that not just social media users' profiles but the whole system of democracy is at stake. This could be the official end of modern times. The power to be able use of big data to manipulate elections shed light on frightening scenarios that might happen sooner than no one has expected. That is why, we must get used to asking 5 Ws and 1 H questions. It won't be an easy task but at least we must try not to be puppets manipulated by the big data. To conclude we shouldn't be afraid of asking big questions no news sources.

Our brain, which is just made of a soft tissue, is home to approximately 86 billion neurons. Compared to number of stars in the Milky Way, the brain can be assumed as a small galaxy. Every activity in the brain is produced by interconnections of neurons and their release of neurotransmitters in response to nerve impulses. Like neurons, Central Processing Unit (CPU) in a personal computer has transistors. Each neuron acts like a single circuit, like a computer transistor, but of course they are not the same. Almost every CPU needs a cooling system that prevents overheating so-called stress. If a CPU overheats, that might cause malfunction or system crash because transistors can't run under stress. Like a CPU, the brain might not run properly under stress. Recent studies showed that chronic stress has a shrinking effect on the area of the brain responsible for memory and learning. While stress can shrink the brain, it can also increase the size of the amygdala, which can make the brain more welcoming to stress. To comprehend why stress influences the brain, it's imperative to comprehend a little about how the brain functions. The brain isn't only a solitary unit, however a gathering of various parts that perform various assignments. On the off chance that you are in a hazardous or genuinely troublesome circumstance, the amygdala (the piece of your cerebrum that administers your endurance senses) may assume control over, leaving the pieces of your mind that help to store recollections and perform higher-request errands with less energy. Since the brain doesn't have a cooling system, we need to avoid stress as much as possible.

To get rid of stress for mental health, there are things to do to protect ourselves: To more readily adapt to stress, we need to consider how we may limit factors that make it worse. To change an attitude towards something is always a challenge but we need to try to change our mindset towards stress. In modern days, it is almost impossible to think of life without stress. Even, limited quantities of stress might be wanted, helpful, and even solid. Experts says that positive stress improves athletic performance. That is why, rather than seeking a life without stress, we had better develop responses to stress.

We are surrounded by technological devices everywhere. Especially screens are the ones that might harm our eyes in the long run. Screens causes digital eyestrain which means that blue light from digital devices might decrease contrast leading to eyestrain. According to some research papers, exposure to blue light for a long time might cause retina problems that could lead to macular degeneration. Apart from these, blue light affects our sleeping habits in a bad way, and it takes us longer to fall asleep. So, what we are trying to underline is the fact that sleep is crucial to our health as eating. With the help of the sleep, body especially the brain repairs itself. Poor sleeping habit caused by blue light is linked to problems as mental health problems such as depression and anxiety. At the end, if we desire to keep our mental health, we should be careful about stress, blue light that might harm our mental health in a way. All that means that we need to be unplugged on a regular basis. Knowing everything doesn't bring us happiness, furthermore it makes us more stressful.

THE REASONS WHY LIFELONG LEARNING IS IMPORTANT AND HOW TO BECOME A LIFELONG LEARNER

Modern times has proven that life span of professional information has become limited to a few years. That leaves us with the feeling that the education we receive might not be enough sooner than we think. Therefore, becoming a lifelong learner is a must if we ever want to maintain success. Living in ever-changing world, one must adapt to new trends in technology. On the other hand, almost every 18 or so months computer processing speed doubles known as Moore law(Hammond, 2011). Of course as humans we can't evolve as fast as CPUs but we need to be aware of what technology has become capable of.

Only the people who can take on the habit of becoming a lifelong learner can cope with definitive punishment so called dilemmas of technology. Lifelong learners read consistently and efficiently. Reading is an incredible method to open up new horizons, train your cerebrum and reform your life. Literal texts whether they are digital or not totally change the manners we see the world. Reading is simply a consistent improvement. Through reading, you can interact with the wise from the past and gain their experiences they share. Lifelong learners love reading and perform is consistently.

There are unlimited ways to develop ourselves. Regardless of whether it is online or offline, in courses available you can participate without spending a dime on it. This is an extraordinary opportunity to interact with similar individuals. Due to the cutting edge innovation, it is quite easy to pick up information from online projects, you may even learn coding from best universities. What lifelong learners have in common is to use these chances to improve their selves as much as possible. Lifelong learners enjoy encountering steady development. They are aware of amazing change that occurs on their learning cycle. One of the key aspects in lifelong learning is having an objective. To continue developing, you need to define your objectives. Lifelong learners love challenges. Since they are not afraid of troublesome objectives, it keeps them motivated. In short, ambitious objectives increase people's performance. Lifelong learners embrace change because a successful change can yield fantastic outcomes. That is why, lifelong learners believe that it is never late to learn something new. People who can't leave their comfort zone don't tend to become lifelong learners.

Especially nowadays, professions admit the importance of developing practitioners to be lifelong learners. Because of that, Students should be supported in developing the attributes of a lifelong learner. From a lifelong learning view point, formal education is significant when it makes people ready for later learning. Lifelong learning also includes an imperative duty in terms of economics: It supports sustainable development. Lifelong learning has been describes as essential to the accomplishment of sustainable development and quality education. the United Nations (UN) purposely recognises the inner role of lifelong learning in providing sustainable development (UN 2015a). The report underlines the importance of promoting lifelong learning opportunities for all. Quality education and lifelong learning open doors for everyone (Funston, 2010). Those enables productive lives and causes sustainable development.

It is considerable to think that lifelong long learning has a huge potential. In any case, it is not just sufficient to create instructive opportunities for grown-ups. We need to adapt a comprehensive, incorporated way to deal with lifelong learning strategy and planning must be in detail. We should create associations and collaboration that cross the limits between formal, non-formal and informal learning that must include the private sector, civil society and the government. If we can thrive on that we will be able to acknowledge completely the full potential of lifelong learning.

Learning is an individual action dependent on an individual's own advantages and inspirations. Supporting such learning exercises causes it possible for individuals to have solid, sound lives. Additionally, securing and refreshing abilities and information makes it conceivable to have a economically steady life. Simultaneously, this sort of learning movement ought to add to the advancement of a person's capacities. Such people will eventually add to the general galvanization of society and the feasible advancement of their nation.

We should be determined about the idea that the society can only prepare themselves for the difficulties of things to come by turning out to be lifelong learners. In this way, humanity will be ready for the future. There are signs that communities around the world learning about the importance of becoming lifelong learners. Countries are realizing advantages of investment in learning throughout life. We have to expand on the case of understanding being committed to lifelong learning.

RISKS FOR LIFE LONG LEARNERS

Everything in life is subject to uncertainty. But ''uncertainty'' does not really mean equivalent to "risk". The critical factor in changing uncertainties into risks is the way that institutions and individuals determine targets to be met. For the institutions these destinations are the key points of the business, while business targets are characterized in the business. The goals characterize the proportions of achievement, and are generally communicated in fixed quantifiable terms—certain degrees of executional targets. Thinking the dimensional of conceivable future states we may envision as could reasonably be expected, the destinations portray our ideal future, our objective or mission the end to which our efforts are reaching. In short, risk can be explained as uncertainty that matters.

Lifelong learning can be characterized as nonstop learning endeavours at any phase of daily routine that assist individuals with improving their experiences and adjust to society. Like everything else in life, there are many risks for lifelong learners to be managed. Supporting lifelong learning is an activity that can be coordinated by the state. Every state needs to be aware of the risks in lifelong learning activities. For society, updating its citizens' knowledge should be managed within lifelong learning concept. This concepts involves a difficult path which is full of uncertainties. Therefore risk management is one of the key factors that must embraced within lifelong learning concept. Modern states wish their societies to maintain a life in welfare. They take care of risks insuring their citizens against hazardous situations. They shield them from social mischief and financial calamity, and controlling monetary dangers and natural perils.

Development of new knowledge through lifelong learning is an idea behind "risk society" and it requires a risk management strategy (Beck,1992). Lifelong learning programmes contain a wide range of instructive exercises. These exercises might be carried out at schools, organizations, public venues, libraries, galleries or enterprises. Risk taking practices in lifelong learning that include dynamic circumstances in which the alternative decisions are described by a "absence of certainty and the possibility of failure. Failure is burden that lifelong learners cannot bear. Moreover, participating in adaptive learning activities (sharing speculative thoughts, posing inquiries, endeavouring to do and learn new things) makes the situation for lifelong learners more complicated. Learners take the risk of making mistakes that might give them the feeling that they are less skilful than others. In other words, people who take part in lifelong learning activities take the risk of making mistakes. Especially for adult learners, if they feel that their ideas are dismissed or ridiculed, that prevents them taking part in learning activities. Although creating a risk free environment is almost impossible in lifelong learning environments, learners must be well informed that taking the risks in lifelong learning activities with all expected advantages will outweigh the negative sides.

There are a variety of individual and relevant components that impact an individual's ability to face academic risks. That is to say, they include interest at the topic, self-efficacy, accomplishment of objectives. For instance, people might be bound to face challenges in subjects in which they are intrigued. A well-developed and stable personal interest in a subject may support risk taking more than situational interest Self-efficacy means that an individual has trust in his capacity to effectively achieve a goal. High self-efficacy —especially creative self-efficacy efficacy (i.e., students' beliefs in their abilities to generate new and meaningful ideas)— can give an individual the certainty expected to participate in risky lifelong learning activities. People with solid self-efficacy beliefs are more ready to see chances as they are additionally more willing to really face challenges, and they frequently continue learning when troubles present themselves. Studies demonstrate that self-efficacy is more strongly related to the willingness to take intellectual risks than is actual ability. all in all, for supporting scholarly danger taking, seen capability might be a higher priority than real expertise level. In other words, for supporting risk taking, perceived competence may be more important than actual skill level in lifelong learning.

In addition to assumptions about self-efficacy, there are other issues that are related to risks such as motivation. Motivation indicates all the arguments that control willing behaviour. In lifelong learning, it is like an engine to keep learning moving forward. Motivation is affected by an individual's discernments, qualities, interests, and objectives. It is always at stake - easy to lose; difficult to gain. Because of that, it is always at risk. For lifelong learners, motivation is a pigeon ready to take off with movement of a heartbeat.

Risk management in lifelong learning process relies upon an individual's emotional view of risk. To put it another way, what a learner understands from risk can be different from someone else's perception. For instance, someone who has detailed information about engines can easily share his information with others. On the other hand, someone who has minimal background knowledge can be too hesitant to share his thoughts. Additionally, if an individual is talented at a task, it will be not too risky to undertake that task in a lifelong learning environment. Another key point to remember is that taking risks in lifelong learning process can be tricky.

Keeping in mind that motivation, self-efficacy, willingness are key factors for lifelong learners. One should also bear in mind that environmental factors can affect the key factors aforementioned like a person's willingness to take risks in lifelong learning process. Lifelong long learners tend to manage the risk regarding their learning experiences, not to mention students are less likely to take risk in classroom based learning environments. Of course, facilitator's' role in supporting students' approach to learning can't be compared to the concept in lifelong learning systems since concept of supporting the learner is completely based on a system that must be managed within risk management approach. Providing this type of support involves "creating a safe atmosphere in which lifelong learners feel comfortable to explore, experiment, and take risks in problem solving and learning." Above all, it seems pertinent to state that lifelong learning process can only be managed by risk intelligent intuitions that take on the principle "quality comes first".

IMPORTANCE OF RISK INTELLIGENCE

Literacy is the lifeblood of ideal citizenship in a democratic society. But nowadays it is not enough just to have literacy skills. Risk literacy is the necessary skill to exist in modern society (Gigerenzer, 2015). Risk literacy is now encountered even in everyday life. For example, new weather technologies have allowed the use of numerical precision expressions instead of verbal expressions in weather forecasts. However, the use of numerical expressions in the prediction did not allow the message to be fully understood (Gigerenzer, 2015). Making life complicated doesn't make things easier. Overly complex systems are difficult to understand. Such systems are easy to abuse and do not give trust to the public. Simple rules help us create a safer world. The simplification of life is essential for the meaning and enlightenment of human life.

Enlightenment is always a phenomenon of future for Kant. Enlightenment is the emancipation of the immaturity that man imposed on him. Immature is the inability of a person to use his own mind without the guidance of another. The reason for this is not lack of understanding, but indecision and lack of courage. Freedom of expression, right to vote and protection from harm are the greatest rights acquired throughout human history (Kant, 2019). The uncertainties brought with modern times are a threat to individuals and institutions. In order to be prepared against this threat, it is thought that individuals and institutions should be equipped, that is, they should have risk intelligence as a risk literate.

Risk intelligence requires some intuitive psychology and some statistical knowledge. It is possible to take control of our lives with curiosity and courage together with skill (Gigerenzer, 2015). Risk intelligence passes through the ability of an individual or organization to efficiently weigh risks. It includes classifying and calculating risks. Understanding relationships is the ability to store relevant information and act on it and adapt to new conditions (Apgar, 2006) How can we decide when to be cautious and when to be brave? How can we be more successful in evaluating our personal risks (Evans, 2012). Two things are required to gain a deep understanding of a risk. The first is to find answers for possible solutions to problems caused by risk; The second is the experience to show the correct solutions and which solutions are wrong. Apgar (2006) places experience on the basis of risk intelligence. For him, experience can help us solve problems that require an understanding of risk.

There are four rules of Risk Intelligence:

1- To learn all the risks that can be learned: Risk management has an important place in the modern economic structure. Most of the risks that businesses have to manage are in the financial risk group, and most of these risks are in the learnable group. For example, the information about how many equipment the agricultural spraying company wants to sell will be sold is very important for the business. An incorrect estimate to be made in this regard may put the company in a difficult situation. However, this uncertainty can be eliminated if the potential customer portfolio is analysed sufficiently.
2- Identify the risks that you can learn the fastest: Just as we cannot choose our parents, we cannot determine who our children will be. However, we are free to choose the risks we bear. It is possible to gain the necessary experience for risk intelligence by learning the risks that we will make a difference in the fastest way.
3- Do the risky tasks in order: Running risky businesses at the same time reduces the success rate. For this reason, it is recommended to put the risky tasks in order and complete them.
4- A business network of stakeholders should be maintained to manage all risks: Business partners whom we are working with should be reviewed. The risks encountered can be shared according to the tasks performed in the same common network (Apgar, 2006).

Risk intelligence is not actually an innate talent or an intellectual skill. It is always possible to develop it as it is a skill based on continuous observation, discovery, and learning. Risk intelligence is not limited to individuals. It contains all the experiential and information resources that we can use to find solutions to a risk problem. Risk intelligence can therefore be named as the basic competence of the individual, team or organization for sustainability (Apgar, 2006; Funston, 2010). At the core of risk intelligence is the ability to measure the limits of your knowledge - being cautious when we don't know too much, and being confident when you know a lot. People with high risk intelligence do just that (Evans, 2012).

Risk intelligence is considered to be significantly improvable under the right conditions. Risk intelligence is the ability to accurately predict probabilities. Risk intelligence is a unique mean found in the middle of overconfidence and insecurity (Evans, 2012). David Apgar defines risk intelligence as "the ability to make the right decisions about a particular new risk." According to him, people seeking different information sources are likely to have higher risk intelligence than those with a lower cognitive horizon (Apgar, 2006).

According to Funston (2010), risk intelligence is "the ability to effectively distinguish between two types of risks: risks that must be avoided in order to survive by preventing loss or damage; and the risks to be caught by gaining competitive advantage". This approach to risk intelligence highlights a common tendency among operators to think of risk in an entirely negative way; this is something that should always be avoided or at least reduced. Funston calls attention to the issues with this view and encourages managers to consider opportunities and threats.

Evans (2012) sees risk intelligence as the cognitive capacity of the individual's brain. It is not enough to have well-designed risk management policies and procedures to build a risk-minded organization; People who implement these policies should also have individual risk intelligence. Risk intelligence is a cognitive capacity and it is purely an intellectual skill. Sometimes risk intelligence and risk appetite get confused. Risk appetite is an emotional feature. It's about how comfortable people are with taking risks.

Institutions with risk intelligence work on data that need to be monitored, analysed and enriched with human insights. The emerging phenomenon of risk intelligence should then be adopted by all staff at the right time. Institutions with risk intelligence aim to develop and adapt as the risk environment changes. In this framework, managers should seek new opportunities and transform them in a way that ensures sustainability in the organization. Organizations that see corporate risk management as preventing loss as well as creating value can optimize long-term performance thanks to risk intelligence (Deloitte, 2010).

It is thought that Institutional Risk Intelligence consists of three pillars. Accordingly, risk intelligence is an institutional formation where risk awareness, risk culture and risk literacy are together. Building a structure that includes the dimensions of culture, awareness and literacy is thought to be a long process that takes time and requires in-service training.

RISK INTELLIGENT COMMUNITIES

For the education industry, distance education plays an important role to transform society that will suit the next century. It enables innovation and ınnovation happens in every section of the industry as technology develops. Innovation creates opportunities for progress leading to improvements in liveability for society. On the other hand, rain of data pours over the society causing confusing times. While there is an avalanche of new data emerging to drive new insights leading to innovation, what is lacking is the risk intelligent communities that can define their priorities. In a risk intelligent community, lifelong learners can use their learning resources effectively. In this case, Technology is the enabler for innovation through data for managing the risks.

risk intelligent community comprises people who are risk literate and aware. It adopts legitimacy based on risk culture. Risk intelligent communities are bounded by responsibility. Only individuals who take up the responsibility with their own learning can be lifelong learners. In a risk intelligent community, every member help each other out. Communities based on trust build up a nation strong enough to survive in post-truth era. Similar minded people like lifelong learners associate with similar, like minded communities to form larger communities. That is why, starting with a small community plays an important role

A Risk intelligent community understands and manages its risks. It is focused on making their community risk aware. Lifelong learning environments are the suitable places to build up risk intelligent communities. A society that can manage its own risk will also be a significant fiscal power in transforming the way cities and communities function. What will make a community risk intelligent will be determined by its investment in social capital and technology infrastructure, coupled with deep educational effort that will continue at all costs, within the ecosystem of communities they are connected around the World

CONCLUSION

It is a challenging environment for public in this Covid-19 times. This give policy makers an opportunity to make things solid for the people. Society may survive intellectually if it keeps its investment on education. Distance education at this time opens a door for everyone. While working on that, a top-down approach might not help people. Within traditional governance structures established around multiple layers of competence where lines of accountability and responsibility are not clear. For the future to come any time soon, we need risk intelligent communities that consist of lifelong learners.

REFERENCES

Apgar, D. (2006). *Risk intelligence: Learning to manage what we don't know*. Harvard Business Press.

Beck, U., Lash, S., & Wynne, B. (1992). *Risk society: Towards a new modernity* (Vol. 17). Sage.

Beghetto, R. A. (2009). Correlates of intellectual risk taking in elementary school science. *Journal of Research in Science Teaching: The Official Journal of the National Association for Research in Science Teaching*, *46*(2), 210–223. doi:10.1002/tea.20270

Britannica. (2020). *Biography of John Ray*. Retrieved 25 September 2020 from https://www.britannica.com/biography/John-Ray-English-naturalist

DePrince, A. E. III, & Hammond, J. R. (2011). Coupled cluster theory on graphics processing units I. The coupled cluster doubles method. *Journal of Chemical Theory and Computation*, *7*(5), 1287–1295. doi:10.1021/ct100584w PMID:26610123

Evans, D. (2015). *Risk intelligence: How to live with uncertainty*. Simon and Schuster.

Funston, F., & Wagner, S. (2010). *Surviving and thriving in uncertainty: Creating the risk intelligent enterprise*. John Wiley & Sons.

Gigerenzer, G. (2015). *Risk savvy: How to make good decisions*. Penguin.

Gordon, J., Halász, G., Krawczyk, M., Leney, T., Michel, A., Pepper, D., ... & Wiśniewski, J. (2009). Key competences in Europe: Opening doors for lifelong learners across the school curriculum and teacher education. *CASE Network Reports*, (87).

Kant, I. (2019). *Answering the question: What is enlightenment?* Strelbytskyy Multimedia Publishing.

McIntyreL. (2018). *Post-truth*. MIT Press.

Ohanian, S. (2008). In a Time of Universal Deceit, Telling the Truth Is a Revolutionary Act. *What is Authentic Educational Reform? Pushing Against the Compassionate Conservative Agenda*, 199.

UN (United Nations). (2015a). *Transforming our world: The 2030 Agenda for sustainable development.* New York: United Nations. Retrieved 25 February 2019 from https://sustainabledevelopment.un.org/ post2015/transformingourworld

Chapter 2
Design Thinking for Online Learning:
A Design Thinking Approach

Soner Sözler
Zonguldak Bülent Ecevit University, Turkey

ABSTRACT

With the spread of COVID-19, nearly all education systems in the world pivoted to distance education in order to overcome the crisis. Because it is an unexpected crisis, most institutions had to employ emergency remote teaching systems. This decision brought so many problems to education industry. But now, it is a known fact that the virus is with us at least one more year, and moreover, new epidemics may occur. Therefore, this chapter aims to provide a guideline with a different approach to online learning for practitioners working in education industry. Thanks to this guideline, the practitioners can create well-designed online learning environments, and as a result, they will not need emergency remote teaching applications anymore.

INTRODUCTION

The world is experiencing a crisis affecting all sectors including education. Especially, schools that have bigger campuses and more students have no possibility to continue main stream education any more. UNESCO, May 2020, reported that covid-19 affected more than 990 million learners and furthermore there are school closures in 130 countries (UNESCO, 2020). Under these circumstances, the only cure for providing educational services was of course Distance Education. The institutions that have DE experience or a well-developed DE system had nearly no problem in

DOI: 10.4018/978-1-7998-4534-8.ch002

pivoting to DE. But others failed to provide service to learners. It is stated that most schools provide asynchronous and self-paced online learning services (Burns, 2020). This has so many reasons such as teachers' having little familiarity with the required technology, access to online resources (digital divide), insufficient infrastructure, lack of content and so on (Insee, 2020), (News, 2020).

Moreover, even in developed countries it is possible to see bad examples in pivoting to DE. UNESCO reports that, in USA. 43% of the teachers are prepared to stimulate remote learning and just 1 in 5 of the teachers said that school authorities furnished guidance (Duupdates, 2020).This situation is similar and even worse in other countries, the report says. As a result, bad examples of OL, which gives harm to the field, have been created.

The situation mentioned above bring the term Emergency Remote Teaching (ERT) into the spot. This term has been defined as a temporary shift of instructional delivery to an alternate delivery mode due to crisis circumstances (Hodges, Moore, Lockee, Trust, & Bond, 2020). It is also defined as an alternative and unplanned method for delivering instruction from a distance because they are simply not able to be physically located in a classroom with their students (Sishley, 2020). So, it can be summarized that it is an approach that has time limit for preparation, unplanned and temporary.

On the other side Online Learning is defined as synchronous or asynchronous telecommunication activities in which students have their own premises (Yang, 2020). When the definitions compared, it can be seen that they have many different features. However, the most important difference is time. Because of time issues, ERT cannot be planned in a detailed way but OL needs substantial amount of time to be planned well. Table 1 shows some different features of ERT from OL.

In addition to these ERT issues, this pandemic makes people in the education sector think about the future of the schools. Neil Fassina Ph.D., president of the International Council for Open and Distance Education (ICDE), posits that rather than a temporary lift, we are in fact at the start of a permanent "long-term uptick, where digitally-enabled learning [will] become a quality, viable and first choice solution." In other words, it is stated that the digital learning isn't a substitute for traditional education—it's the next frontier (Rosen, 2020).

However, because of having a world-wide impact, this temporary approach and discussions about the future of education caused a permanent idea of a well-designed OL environment. It is clearly seen that; without a systematic approach it is not possible to provide a well-developed OL environment. Although there are numerous alternatives to design an OL environment, this chapter presents a unique way of designing an OL environment.

Table 1. Emergency Remote Teaching and Online Learning

	Emergency Remote Teaching	*Online Learning*
Delivery of Instruction	Substitute for face-to-face learning because of circumstances.	Online is always envisioned as the primary means for the delivery of instruction.
Preparation	A quick consolidation of a large amount of course material prepared for face-to-face learning into an online format. This transformation results in improvised solutions to accommodate a new modality of learning.	Course design requires a substantial amount of time, forethought, and preparation before the course begins.
Integration of Technology	Limited technology may result in less than ideal situations of engagement and interaction.	The purposeful integration of technology in the course to create an engaging and interactive experience for learners.
Technology Training	Faculty experience a quick adaptation of technology for the delivery of online instruction with minimal professional development. This quick adaptation may result in a stressful situation for face-to-face faculty.	Faculty professional development is planned and focused. Faculty are trained weeks if not months in advance in the technology used for teaching online.
Online Pedagogy	Faculty who teach face-to-face courses typically do not receive training in online teaching pedagogy.	Faculty are trained in online pedagogy prior to the movement of courses into an online format.
Course Design	Courses are designed for face-to-face delivery.	Courses are designed according to instructional design principles that concentrate on online learning.

Source: (Sishley, 2020)

BACKGROUND

"What is" Design Thinking?

In terms of history of Design Thinking Brown (2009)states that DT is, as a mindset and methodology, relatively young. In comparison, the scientific method has stood centuries of rigorous investigation; and modern management practices such as Six Sigma and lean manufacturing have benefited from decades of practice and examination. Design thinking has seen just 15 or so years of widespread adoption. For the most part, it's still largely a set of heuristics for guiding team-based collaboration.

When it comes to its definition, the term Design Thinking was first used in the 1990's by David Kelley and Tim Brown of IDEO, with Roger Martin, and encapsulated methods and ideas that have been brewing for years into a single unified concept. Tim Brown, Executive CEO of IDEO, 2012, defines DT as a human-centered approach to innovation that draws from the designer's toolkit to integrate the needs of people, the possibilities of technology, and the requirements for business success. In addition to this definition, IDEO adds that DT is an idea, a strategy, a method, and a way of seeing the world. It's grown beyond the confines of any individual person, organization or website. Brown and IDEO shapes DT with its 3 core activities which are ideation, inspiration and implementation.

Figure 1. 3 core activities ofDesign Thinking, by IDEO, 2012, https://designthinking.ideo.com/. Copyright 2012 by IDEO.

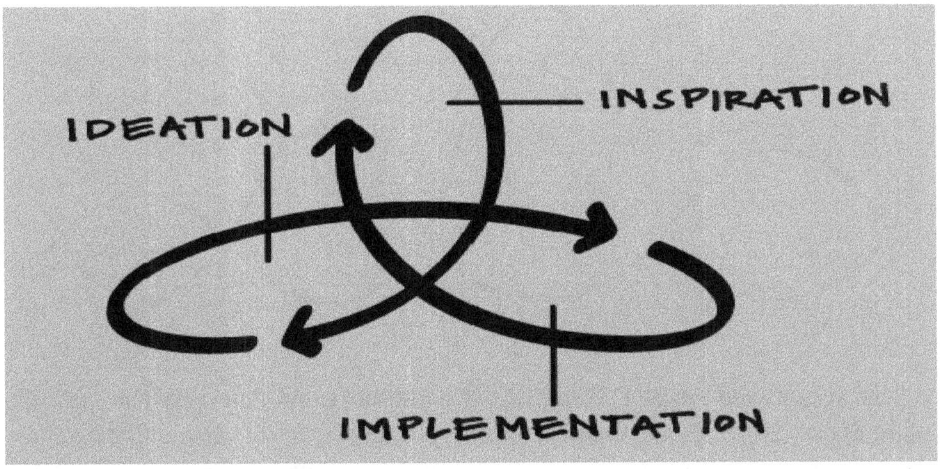

This model of DT can be called the first model that has a systematic approach to problems. Although it has different activities in the sub-level, it focuses on these 3 core activities basically. Thanks to this, this model can be employed in more than one field. It can also be said that all 3 core activities are iterative. That is, in any stage of the model, it is possible to turn back and improve the approach and produce a better solution. Thanks to this iterative structure, DT successes to stay updated and in the center of the issues.

Liedtka (2014) sees the term as a result of a team effort. Therefore, she adds that one can hardly win alone. Expanding on Liedtka's view, design thinking is the collective effort of a few individuals working toward a common goal. Design thinking's collaborative methodology impacts teams in a way that the focus shifts from "I" to "we" and the group's attention is redirected to what matters: the end user's problem. According to Liedtka (2014), the design thinking process needs to answer four questions — What is? What if? What wows? What works? — each representing a different stage of a design experience. "What is" refers to the present. "What if" builds on the present to envision a new future. "What wows" helps teams focus on solutions that stand out and "What works" tests possible solutions with actual users in the real world.

Figure 2. The four questions of Design Thinking, by Jeanne Liedtka, Andrew King and Kevin Bennett, 2013, https://static1.squarespace.com/static/590a5acf15d5dba8afd18da5/t/594c93a56b8f5bb71960c193/1498190757881/S%26L+10+stories.pdf. Copyright 2013 Columbia University Press.

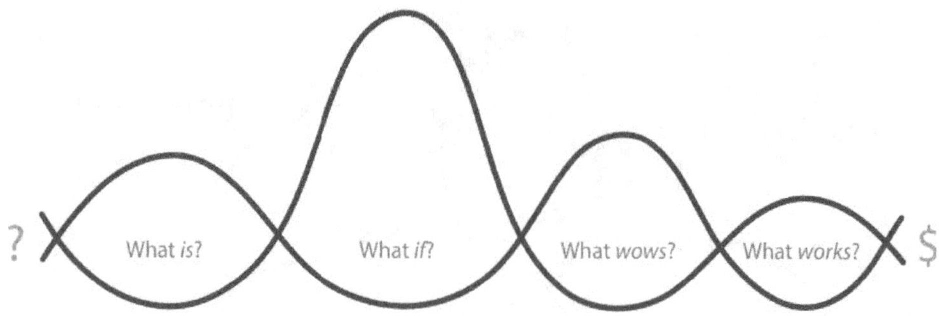

This second version of DT employs 4 questions to produce problem specific solutions. It starts with a "What is" question symbolizing the deep understanding of the problem and then it moves to the next question. Although it seems like a step forward approach, it has an iterative structure, as well. As an example, when you ask the last question which means to test the idea, it is possible to turn back the first question to check the problem. As one can understand, each question represents a stage of DT. Similar to the first model of DT, this model can also be applied in different fields as well as education. As a different feature, this model emphasizes the team work and it claims working collectively can take you to your common goal with better solutions.

British Design Council (2015) has also contributed to the definition of DT and they describe it as a process of exploring an issue more widely or deeply (divergent thinking) and then taking focused action (convergent thinking). They added that each design may have a different point of view but in the creativity processes they have common points. Then they define these common points as 4 Ds: Discover, Define, Develop and Deliver. They improved a model called The Double Diamond model uses these 4Ds of design in an iterative way.

Figure 3. Double Diamond Model, by British Design Council, 2015, https://www.designcouncil.org.uk/news-opinion/what-framework-innovation-design-councils-evolved-double-diamond. Copyright 2015 by British Design Council.

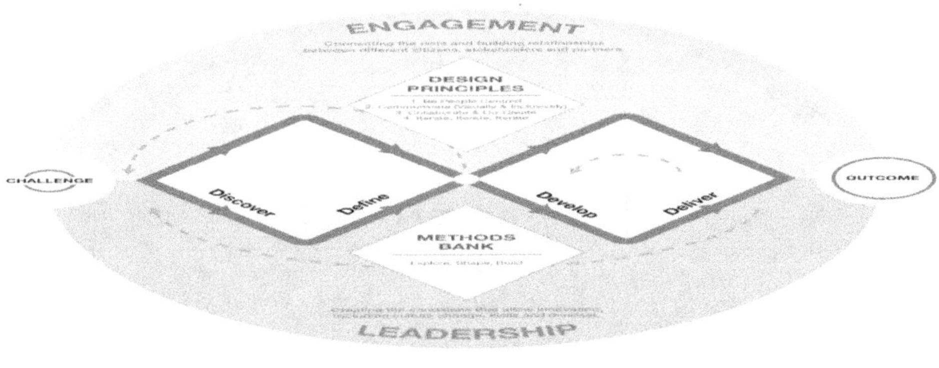

This model developed by British Design Council based on the idea of exploring an issue deeply. While doing this, this model uses 4 stages effectively. Similar to previous models, this model is also iterative and each stage of the model has different sub-activities that help the model produce answers for each stage. In any of the stage, it is possible to turn back to the previous step to reach the outcome with a better understanding of the problem. This model also emphasizes being human-centered feature of DT as a principle. Moreover, sees the problem as a challenge to overcome and like other models it can also be used for large-scale problems as well as small scale problems.

The models given above are close to Gibbons' (2016) who defines DT as an ideology supported by an accompanying process. She gives the full definition as an ideology that asserts a hands-on, user-centric approach to problem solving, can lead to innovation, and innovation can lead to differentiation and a competitive advantage (Gibbons, 2016). This hands-on, user-centric approach is defined by the design thinking process and comprises 6 distinct phases, as illustrated below.

This DT model uses innovation as a starting point and it offers collaborative work to produce practical solutions to problems encountered. Like the all models of DT, it has an iterative structure and 3 main stages that have 6 sub-stages. This model can be preferred for larger-scale problems because of having very detailed stages and activities to reach out different problems. It also needs team-work as in two other models mentioned above.

Figure 4. 3 buckets and 6 phases of Design Thinking, by S. Gibbons, 2016, https://www.nngroup.com/articles/design-thinking/. Copyright 2016 by S. Gibbons.

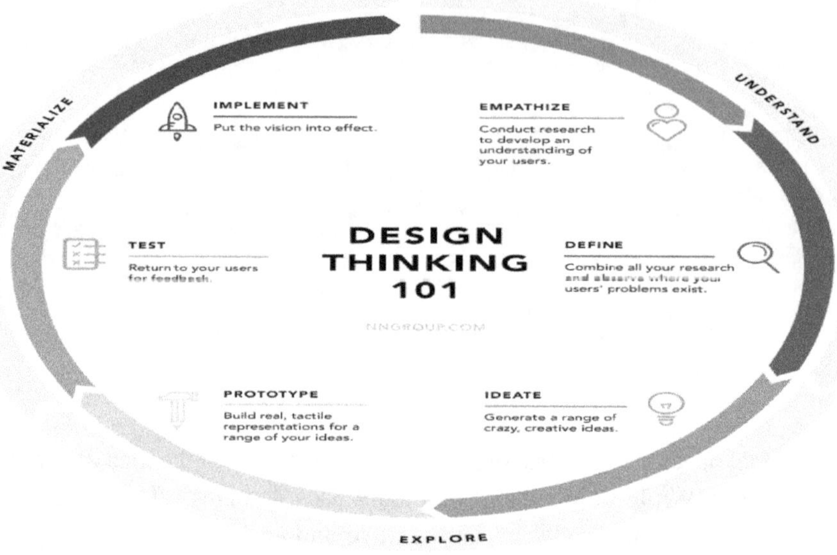

Figure 5. Evolution6² Mindshake Design Thinking Model, by Tschimmel, et al., 2018, http://www.d-think.eu/uploads/1/6/2/1/16214540/researchreport_d-think-dv.pdf. Copyright 2018 by Creative Commons.

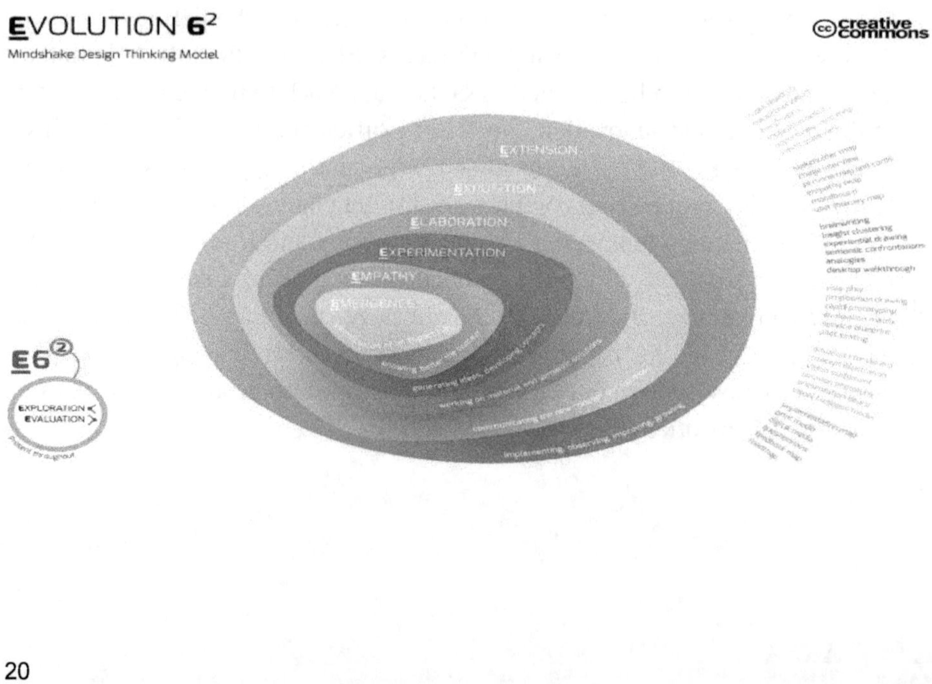

In a project called D-Think, Tschimmel and her friends (2018) gives a detailed definition of DT. They expressed it as a method and a process for investigating open and ill-defined problems, acquiring and analyzing information, identifying opportunities for innovation, deepening empathy, experimenting with new perspectives and visualizing new concepts. Also they added that it is suitable for education because of its features.

This very current model of DT has so many toolkits that help people to understand the problem better. Besides being an iterative model, its each stage covers one another that shows the model as a whole. As given in its definition, it sees DT as a process and method which needs uses a systematic approach to produce alternatives for the ill-defined problems.

There is no doubt that the definitions of DT will continue to evolve but it can be summarized that DT is a mindset that has solution-focused steps. It can also be thought as a process, an approach or even a method. Moreover, its implementation area is endless and it can easily produce practical solutions for the problems.

In the light of the issues mentioned about ERT and OL, it is clear that OL needs detailed planning. With a well-designed OL environment, it is easy to create face-to face teaching effects on the shareholders of the education field. In addition to these effects of OL, with a well-designed OL environment it is possible to use the advantages of OL that a traditional classroom cannot provide. Furthermore, because of the sudden shift to OL nearly all members of the education era are being asked extraordinary things regarding course delivery. Therefore, DT can help them overcome this extraordinary burden.

A DESIGN THINKING BASED ONLINE LEARNING ENVIRONMENT

Step 1: Understand/Discover the Problem/Challenge

This chapter gives 5 current models of DT. All models given here starts with understanding the problem deeply. Before designing any OL environment, you should use any resource to know the target participants (learners) better. So, you can follow more than one way to reach out this aim. You may list the problems, you may want your participant to write a brief, you may use a brainstorming activity, you may employ a questionnaire or make observation. Because DT has a clear nature, try to keep step 1 as simple as you can.

One of the methods that you can employ in the first stage of DT is dreams and gripe session. In this method, you can simply ask a question about the wishes or dreams that the participant hopes to exist in the learning environment. Figure 6 gives you an example of dreams and gripe session of step 1.

Figure 6. Dreams and Gripe Session, by IDEO,2016, https://www.ideo.com/post/design-thinking. Copyright 2016 by IDEO.

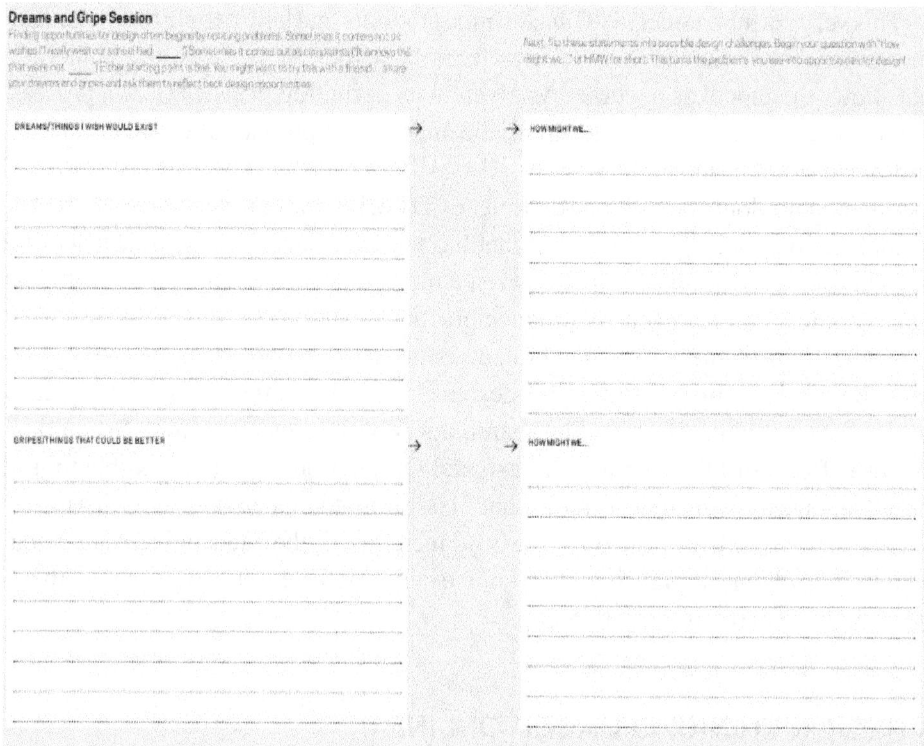

With a very simple list, you can learn the needs of your learners. If you already know your learners well, you can use this list for yourself, as well. If you have a multinational class, it can be a good idea to want them write in their own languages. By this way, they can express themselves better. One another point is being sure about the constraints that you want to define. It is an important aspect of addressing the problem.

As an another alternative, you can do a brainstorming activity to understand your target learners better. If you choose this activity, asking these questions to yourself or even to your learners can really help;

- What do your learners do from day to day?

- What is their aim to get this education?
- What are their challenges for OL?
- What equipment do they have?
- What is the level of their familiarity with OL?

Answers for these questions will make you create a better study plan and thanks to these questions' answers you can start designing an OL environment that addresses more learners/participants. Asking these questions also will make you think in your students' shoes and thanks to this you can produce a tailor made OL environment. As the first stage of DT is understanding your learners/participants, this way of approaching is also a way of emphaticizing with your learners/participants.

DT also gives importance to personalities of your learners/participants. Learning about their personalities will also help understand their context, as well. If you understand your learners'/participants' context better, you can design a better environment that addresses to the needs of your learners/participants. It is also important for implementing the right pedagogy to each learner. To understand the personality, you can employ a learner/participant matrix and you can interview with them. This will make you empathize with them better, too. During the interview, try to use personal questions instead of a set of fixed questions. It is better for finding the common concerns of your learners/participants. Stanford university Design School (d-school) suggests in-person or in-group interview to understand your learners/participants better. They stated that they interview to gain empathy. By interviewing users, you will begin to better understand a person's behaviors, choices, and needs. They suggest interviewing in-person and in pairs so that one person can converse while the other captures (d-school, 2020). They also emphasize the clues given below;

- Ask why
    - Even when you think you know the answer.
- Never say "usually" when asking a question
    - Instead, ask about a specific occurrence. "Tell me about the last time you ____."
- Encourage stories
    - Stories reveal how users think about the world.
- Look for inconsistencies
    - What users say and do can be different. These inconsistencies often hide interesting insights.
- Pay attention to nonverbal cues
    - Be aware of body language and emotions.
- Don't be afraid of silence

 - When you allow for silence, you give users time to reflect on their answers—which may lead to deeper responses.
- Ask questions neutrally and don't suggest answers
 - "What do you think about buying gifts for your spouse?" is better than "Don't you think shopping is great?"

Table 2. List of Design Thinking Empathy Stage methods

Stage	Methods
Understand the Problem/ Empathy	• Assume a beginner's mindset • What? How? Why? • User camera study • Interview (for empathy) • Extreme users • Analogous empathy • Story share-and-capture • Bodystorming • Define problem statement • Observe users (may include think aloud protocols) • Conduct site visits (contextual inquiry) (users doing their tasks at their work places) • Interview users, send questionnaires

Source: (Waloszek, 2012)

It can be easily understood that in order to be able to empathy with your learners/ participants deeply, it is better to try to think in your learners'/participants' mindset. The more information you get; the better study plan you set up.

One another technique that can be helpful to understand your participants deeply is Bodystorming. It is a technique used in the first stage of the design process that takes learner/participant viewpoint into account. Instead of trying to visualize or imagine how a product might be used, bodystorming requires the design team to, in essence, act it out as though the product or process already exists. In this way, it is very similar to that simple childhood game that required a dose of imagination and ''pretend play'' to create a more user-friendly experience (Hendricks, 2020). Thanks to this method, you can learn the experience of the learner/participant about the OL environment that they exposed to.

Whatever activity you choose; you should ask yourself "How do I approach the problem?". When you ask this question to yourself, you start creating realistic objectives for your prospective course design. Moreover, your study plan will be based on these questions. Then you can start working on the ill-defined problems and visualize them better. Table 2 shows some methods that can be used to empathize with your learners/participants.

Table 3. Design Thinking Journey Mapping Table

Learner/Participant	*Experience*		
	Before OL	*During OL*	*After OL*

After getting information about the problems encountered, you should move on to the second phase of step 1 which is called defining the problem. In this phase, your aim is to visualize the problems that are mentioned by your learners/participants and to set up a study plan that will focus on producing solutions for each problem mentioned before. Similarly, there are different methods that can be used in this phase of DT. One of them is "How might we" questions.

"How might we" questions can help to get possible solutions to the problems addressed. Because your participants are experiencing the problems themselves, it is highly possible for them to think a solution for the problem. What is more, to get their ideas can make you define the problem better. A design challenge is the starting point of every design process, and the purpose you will work toward. Framing the right "How might we" question to address your challenge is essential. The question should be broad enough to allow for unexpected possibilities but narrow enough to let you focus. Be careful not to include the answer in the question. Here are a few examples you can choose from, or use as inspiration to come up with a challenge that matters for you.

- How might we engage students more deeply in online learning?
- How might we design our online learning environment to be student centered?
- How might we create a space for peer learning?
- How might we adapt school schedule to online learning?
- How might we support our learners in after class hours?

The d-school of Stanford university also published a how to write "How might we questions" guide. In the guide they suggest the following issues to be taken into consideration before writing a question;

- start with your design challenge and point of view statement.
- break down the larger challenge into smaller actionable bits
- ask questions that open up the solution space.

In this defining phase of DT, the understanding part plays a crucial role. Because each detail that you get in the first phase can help you to find a better solution. One important method for defining the ill-defined problems is journey map. A journey map is a tool to dissect a process into its moving parts to illuminate areas of potential insights. Don't forget the details when gaining empathy and understanding for a user and their experience. Creating a journey map is an excellent way to systematically think about those detailed steps or milestones. Journey maps can be used for your own empathy work, or to communicate your findings to others (d-school, 2020). It is a kind of process map that shows each action as separated each other. By this way, you can see all the details of your learners' routine and can feel the problem they encounter.

Figure 7. 2x2 Matrix, by Gecis,2019, https://uxdesign.cc/2-x-2-decision-matrix-bf17364a4993. Copyright 2019 by Gecis.

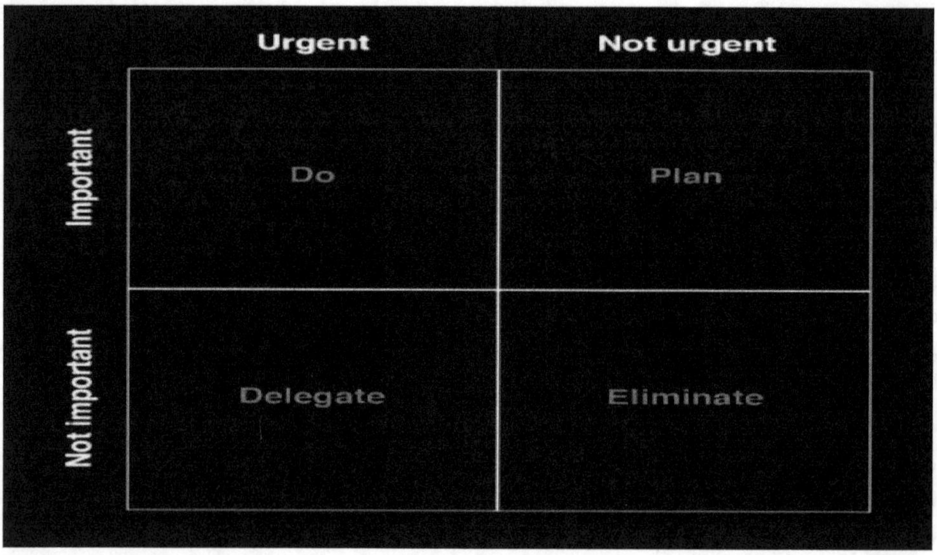

Although there are tools that help journey mapping, just a table can help you, as well. Just think about each learner/participant and use all details you have. Table 3 is a Design Thinking Journey Map sample.

One technique that is used for narrowing the problems mentioned before is 2x2 matrix method. It is just a simple matrix that helps to classify the problems. In this method you can categorize your learners/participants or problems or you can order them into their importance. Just draw an X axis and y axis on a sheet of paper and give a name for each axis. Then start categorizing the matrix according to the information you have. Especially in meetings with your team members, this tool can really help you organize ideas. Also it can show you a starting point. What is important in creating a matrix is to have clear criteria for each axis. Gecis (2019) suggests some tips to have better 2x2 matrix;

- You need tension between attributes (e.g. cost and time, importance and urgency, etc.). People likely choose attributes that are too alike/similar. Those types of matrixes, that plots two alike attributes, won't do the job. Popular 2x2 matrix categories include Cost and Value, Importance and Urgency (the Eisenhower Box), Time and Money, Effort and Impact.
- The best results happen when the team defines the boundary between the quadrants. For example, if the horizontal axis represents the time/urgency it would take to complete a task, the boundary line between the Urgent and Not urgent quadrants might be defined as 1 week, month etc. The same goes for other attributes, like cost, importance, etc.

This sample image not only categorizes the problems but also it shows which problem is more urgent and important. Therefore, you can easily set up your action plan and also you can put your actions in a more correct order. DT has an iterative pattern that enables to update each phase in case of need. So, it will be better to revise your matrix after doing an action. An action that you carried out can also affect the others and hence the place of the items can change. This situation will take you to better results and your OL community will feel better in the end.

One beneficial method for identifying the problems is describing your concept. After a brainstorming or listing activity, you may feel that you don't really understand the idea you selected to act on. In this situation this method can show you what you exactly do during next stages of DT. The idea underlying this method is to select an idea from your brainstorm and flesh out the concept while it's fresh in your team's mind. Asking questions like "*What would you name it? Who is it for? What does it accomplish for them? How does it accomplish this?*" can help you specify the problem that you work for.

As you can easily understand that there are two sub-phases in the first stage of DT. Both phases serve to help the problems understand better. DT's iterative structure makes it possible to go back and forth in the phases. Feeling in your learners'/ participants' shoes will make you design a better OL environment for them. Also, your OL environment will be sustainable, too. You can revise it any time you need thanks to DT. Additionally, you can apply it both for bigger and smaller groups. It also helps your learners/participants question their own needs and progress. By this way, they will be more aware of their own learning which is suggested as an important implication for online learners (Bozkurt, et al., 2020)

As a mindset, DT needs well planning. Before starting your actions, you need to think about all steps that you step on. So unlike to ERT, an OL environment with a DT approach can reach every participant or learner and also can address to each individual separately. Table 4 presents some methods that can be used in the define phase of DT.

Table 4. List of Design Thinking Define Stage methods

Stage	*Methods*
Understand the Problem/ Define	• Space saturate and group • Empathy map • Journey map • Composite character profile • Powers of ten • 2×2 matrix • Why-how ladder • Point-of-view madlib • Point-of-view analogy • Point-of-view want-ad • Critical reading checklist • Design principles • "How might we" questions • Contextual inquiry models (including affinity diagrams, etc.) • Personas, roles • Use cases, scenarios, user stories, day-in-a-life scenarios, etc.

Source: (Waloszek, 2012)

Step 2: Explore/Develop Creative Ideas

This phase of DT is known as generating and prototyping creative ideas. During this phase, the task is to generate ideas that help solve the ill-defined problems mentioned in the previous phase and to prototype a solution for the problems. In this phase, visualizing the problems to is another feature of DT. By this way, you can look with a broader point of view and at the same time you can address to each problem separately. In another words, you've grown to understand your users and their needs in the Empathize stage, and you've analyzed and synthesized your observations in the Define stage, and ended up with a human-centered problem statement. With this solid background, you and your team members can start to "think outside the box" to identify new solutions to the problem statement you've created, and you can start to look for alternative ways of viewing the problem.

One of this alternative ways is focus groups. Known as a qualitative research technique, focus group is also used in usability and user research to collect the opinions of a group of users. It consists of a carefully selected group of participants (between 5 and 10), in a comfortable environment with a skilled moderator (facilitator) who uses a script to ask questions of the group, the answers are recorded and then analyzed and reported to enable further decision making (Foundation, 2020). It can also be thought as a larger scale interview. Moreover, they are helpful for adding meaning and understanding to existing knowledge, or getting at the "why" and "how" of a problem. Therefore, conducting a focus group needs very careful preparation. In a focus group the goal is to generate maximum amount of discussion and opinions. Be sure to;

- Conduct more than one focus group,
    - It makes you check the responses for each question.
- Try to keep the participant number between 5 and 10.
    - More participants can lead to some unheard participants.
- Start with leading and urgent questions.
    - It saves your time to start working on.
- Record all ideas and opinions.
    - It is better to use a chart and a voice recorder.

One other way is storytelling. This method creates an imaginative atmosphere and links your heart and brain. As a result, you feel no obstacles in front of generating creative ideas. It is also said that storytelling is used to get insight into users, build empathy and reach them emotionally. Designers create personas to represent target users and add conflict to stories that reflect their user journeys and problems. Crafting stories, designers can better understand what users want from a solution (Foundation, 2020).

Storytelling is a very powerful method in DT. Brown (2009) commented on storytelling as we mostly rely on stories to put out ideas into context and give them meaning. He adds that stories reinforce the emotional reasons and that is why we remember, and we pass it on to other people. Stories are crucial for brands because they are a powerful tool to make a significant impact. One other comment by Cozma (2019) about storytelling says that astory accompanied by images keeps the listener engaged and curious. Even though we had many ideas along the innovation process, only one idea "stand out from the crowd." That idea had a story with a begging, middle, and end. That is why an idea without narrative structure is doomed to fade away along the design thinking process.

It can be understood that storytelling has a very important place in DT. It is believed that thanks to storytelling people fell as if in an another world and can express themselves better (Brown, 2009), (Cozma, 2019). Therefore, do not forget that your hero in this story is your learners/participants. Let them create their own world and try to reach out every detail you need.

Stoking is another method that can be used in this phase of DT. It is defined as a transitional mode that incorporates kinematics and improvisation in quick team-based activities (d-school, 2020). Stoke activities are synonymous with the term "warm-ups" and they help teams loosen up and become mentally and physically active. Therefore, they can be used when energy is wavering, to wake up in the morning, to launch a meeting, or before a brainstorm (d-school, 2020). Stoking can set the mood specific to any session. DT is a mindset so Stoking can generate a mindset that is specific to any idea.

Stoking serves at least two fundamental purposes:

1. transitioning from the focusing/narrowing mental point-of-view the define/ synthesis phase
   of the design process to the flaring mental point-of-view of ideate
2. invigorating team dynamics

Of course there are more methods to use in this phase of DT. No matter what method you choose, you aim should be to generate creative ideas that you obtained in the previous stage. DT is an on-going process, in case of need you can turn back to any phase that you completed and revise your method in order to produce the better creative idea. Table 5 presents some methods of Explore/Develop stage.

Table 5. List of Design Thinking Explore/Develop Stage methods

Stage	*Methods*
Explore/Develop Creative Ideas	• Powers of ten • Stoke • Brainstorming (+ selection) • Bodystorming • Impose constraints • Focus Groups

Source: (Waloszek, 2012)

Once you generate ideas, it is time to prototype them. This sub-stage is combined with the, explore/develop stage. In this phase, your aim is to build a prototype model of developed ideas. It is a kind of pre-test stage. Before testing any idea, you can check the ideas if they address to the problems or not. Thanks to this checking, you can carry out your last modifications before moving to the next phase. It is kind of evaluating the ideas. This stage is also defined as experimentation and turning ideas into tangible products. A prototype is basically a scaled-down version of the product which incorporates the potential solutions identified in the previous stages. This step is key in putting each solution to the test and highlighting any constraints and flaws. Throughout the prototype stage, the proposed solutions may be accepted, improved, redesigned or rejected depending on how they fare in prototype form (Stevens, 2020).

Table 6. List of Design Thinking Prototype Stage methods

Stage	*Methods*
Prototype	• Bodystorming • Impose constraints • Prototype for empathy • Prototype to test/decide • Impose Constraints • Testing with Users • Identify a variable • User-Driven Prototyping

Source: (Waloszek, 2012)

The prototype phase has so many methods, as well. One of them is prototype to decide/test. This method is used especially for choosing one idea from multiple ideas. If the team is stuck between the ideas and cannot decide which idea to go on, this method can help. In this method, you can test the conflicted ideas with real learners/participants. You list the results and compare them to find the best prototype. By this way, you can decide on the prototype idea that you go forward. As an example, you cannot decide what LMS or telecommunication software to use. In your team, your members set the pros and cons of each. So, you decide to try each with real learners (not a large scale) and compare the results. The results will show you the best software to use for your learners.

One effective method that can be used in this phase is prototype for empathy. You don't have to test all ideas with real learners/participants. If this is the case, this method is very suitable. d-school (2020) defines this method as developing prototypes or create situations specifically designed to gain empathy, without testing a solution at all (or even having a solution in mind). This is sometimes called "active empathy" because you are not an outside observer, you are creating conditions to bring out new information. In the same way a solution prototype helps you gain understanding about your concept, an empathy prototype helps you gain understanding about the design space and people's mindsets about certain issues.

As one can understand that, this phase helps to choose the best opinion/solution. Especially, in case of having multiple problems to be addressed, it is likely to have multiple ideas generated. But before implementing any idea, prototyping it will help real learners'/participants' problems solve easily. One of the main problems with ERT is the quick adaptation of the materials or technology. Using any technology or material without prototyping is a real risk and it can be too late to change the material or technology integrated to your curriculum. Therefore, in an OL environment with a DT approach, it is advised to prototype your materials at least with a pilot group. Table 6 presents some methods of Prototype stage.

Table 7. List of Design Thinking Materialize Stage methods

Stage	*Methods*
Materialize/Deliver/ Implement	• Testing with users • Prototype to decide • Identify a variable • User-driven prototyping • Wizard-of-Oz prototyping • Feedback capture grid/matrix • User tests in the lab (may include think aloud protocols) • Remote user tests • User tests in the field • Informal user tests (e.g. with colleagues, friends, etc.) • KPI studies

Source: (Waloszek, 2012)

Step 3: Materialize/Deliver/Implement Your Ideas

Although it can be seen as the last step of DT, actually there is no end in DT. Because it has an iterative structure, you can improve your ideas after implementing them. You should always think DT as a mindset and plan each step with an aim in your mind. All steps are interdependent and affecting each other. In this third phase of DT, the aim is to materialize your idea and deliver it to the real learners/participants.

Like all phases of DT this phase also has tools that works for presenting your ideas into real world of your learners/participants. One of them is feedback capture matrix. This method is described like giving feedback on progress within the design team or capturing user feedback. The matrix itself arranges thoughts and ideas into four categories for easy assessment. It is very easy to make the matrix. First, section off a blank page or whiteboard into quadrants. Draw a plus sign in the upper left quadrant, a delta sign in the upper right quadrant, a question mark in the lower left quadrant, and a light bulb in the lower right quadrant. Then, fill in the matrix as you give or receive feedback. Place things one likes or finds notable in the upper left (plus sign). Constructive criticism goes in the upper right (delta sign). Questions raised go in the lower left (question mark). And new ideas spurred go in the lower right (light bulb). If you're giving feedback, strive to give input in each quadrant (especially the upper two: "likes" and "wishes") (d-school, 2020).

After validating your ideas and testing your prototypes, you are confident that this solution will impact positively your audience and help them overcome the problems you detected during the research phase. So, this phase will show you if your expectations are true or not. In other words, it is realizing your ideas and it is so important. But it is stated that this phase is one of the most forgotten (Gibbons, 2016). This issue is also not consistent with sustainability. Always keep in mind that DT does not have an end. It is basic aim is to innovate existing products and provide better service for users.

Keeping this aim in mind DT employs identify a variable method in this phase. This method is very time and cost efficient. It can also be used in prototype phase. It is basic aim is to choose a problem specific variable and implement it to see if it works or not. D-school expresses that this method not only saves time and money (you don't need to create all facets of a complex solution), it gives you the opportunity to test multiple prototypes, each varying in the one property. This encourages the user to make nuanced comparisons between prototypes and choose one option over another.

SOLUTIONS AND RECOMMENDATIONS

Online Learning is defined as a term to describe an emerging approach to learn at students' own premise through advanced information-communication technologies (such as Blackboard, Moodle, YouTube, Virtual Reality) either asynchronously or synchronously (Yang, 2020). In this definition, the words "*students' own premise*" is so critical nowadays. Keeping their attention and planning an academic year is also vital. Therefore, this chapter tries to provide a guide for policy makers, practitioners or researchers in the field of OL. Thanks to DT approach, they can design and develop better plans for future. Also it can help people adapt to new normal of education era. It is a common reality that Covid-19 will be surpassed but its effect keeps living in the minds of everybody.

Because of the pandemic, education became one of the most important problem in the world. Nearly all countries cannot develop a permanent solution for this problem. Moreover, most practitioners in education era are alone in designing their own OL environment. Starting with a systematic approach can make them feel better in I situation like this. Furthermore, planning each step of their OL environment will have better results.

Last but not least, ERT can be acceptable in unplanned or emergency situations. But from now on, each individual working for an educational institution is responsible for a well-designed OL environment. DT approach is a way of doing it, but it is also a kind of lifestyle. It is a mindset that enables you to plan each step of your future decisions.

FUTURE RESEARCH DIRECTIONS

Designing OL curriculum with a DT approach would be a good research for learners. The new normal term in this new normal world will be in the center of education in the upcoming years. It would also be great to study on innovating the available OL environments with DT approach.

CONCLUSION

This chapter aims to provide a practical guide for practitioners working in education area. So, it tries to give clear definitions and explanations of the stages and phases of DT. It also explains three common and basic phases of DT approach. In short, Design Thinking can be summarized as an approach that adds innovation to the current problem situation, striving to respond to the needs of the learners/participants it will serve as a starting point and to find solutions to the problem situation and therefore offer different solutions according to each situation. In addition, although different models are exemplified above, it is possible to find common points in terms of thinking. In this context, it can be thought that OL environments also serve different learners/participants in terms of socio-economic, cultural, readiness, age, gender or academic background. If OL environments are designed with Design Thinking, it can be concluded that by seeing each difference as a problem situation, they can produce specific solutions for that situation and thus support learning in a very positive way.

Looking at the framework drawn for Design Thinking approach above, it can be said that OL environments and Design Thinking approach, which takes the participant, namely the target learner, as the starting point, can work in harmony thanks to their dynamic structures and their understanding of innovation and change. Therefore, this approach will provide many conveniences to learners, experts and institutions by using it in the design of OL environments. In addition, it will be able to respond in depth to the needs that may be required in educational environments. In this way, DT will contribute positively to the development of OL environments, providing equal opportunities, managing their own learning and meeting the needs of future generations.

REFERENCES

Bozkurt, A., Insung, J., Xiao, J., Vladimirschi, V., Schuwer, R., Egorov, G., & Paskevicius, M. (2020, June 5). A global outlook to the interruption of education due to COVID-19 pandemic: Navigating in a time of uncertainty and crisis. *Asian Journal of Distance Education*, 1-126.

Brown, T. (2009). *Change by Design: How Design Thinking Transforms Organizations and Inspires Innovation*. HarperCollins Publishers.

Burns, M. (2020). *eLearning Industry*. We Are All Distance Learners: Online Education Across The Globe—Part 2: https://elearningindustry.com/distance-learners-online-education-across-the-globe-part-2

Council, B. D. (2015). *What is the framework for innovation?...* Design Council's evolved Double Diamond: https://www.designcouncil.org.uk/news-opinion/what-framework-innovation-design-councils-evolved-double-diamond

Cozma, A. (2019). *Sidlaurea*. Why is Storytelling at the core of Design Thinking?: https://sidlaurea.com/2019/09/22/why-storytelling-is-at-the-core-of-design-thinking/

d-school. (2020). *Product Discovery Methods*. Interview for Empathy: https://pdmethods.com/interview-for-empathy/

Duupdates. (2020, June 24). *Duupdates*. UNESCO report 2020: Online education implemented during covid-19 are not inclusive: https://www.duupdates.in/unesco-report-2020-online-education-implemented-during-covid-19-are-not-inclusive/

Foundation, I. D. (2020). *Focus Groups*. What are Focus Groups: https://www.interaction-design.org/literature/topics/focus-groups

Gecis, Z. (2019). *UX Collective*. The 2x2 matrix: how to be more systematic about the decisions you make: https://uxdesign.cc/2-x-2-decision-matrix-bf17364a4993

Gibbons, S. (2016). *Nielsen Norman Group*. Design Thinking 101: https://www.nngroup.com/articles/design-thinking/

Hendricks, B. (2020). *Study.com*. Bodystorming in Design Thinking: Definition, Purpose & Example: https://study.com/academy/lesson/bodystorming-in-design-thinking-definition-purpose-example.html

Hodges, C., Moore, S., Lockee, B., Trust, T., & Bond, A. (2020, March 27). *Educausereview*. Retrieved from The Difference Between Emergency Remote Teaching and Online Learning: https://er.educause.edu/articles/2020/3/the-difference-between-emergency-remote-teaching-and-online-learning#:~:text=In%20contrast%20to%20experiences%20that,mode%20due%20to%20crisis%20c-ircumstances

Insee. (2020). *Insee*. Statistics and studies: https://www.insee.fr/en/statistiques

Liedtka, J. (2014). Innovative ways companies are using. *Strategy and Leadership*, *42*(2), 40–45. doi:10.1108/SL-01-2014-0004

News, U. (2020). *Global perspective*. Startling disparities in digital learning emerge as COVID-19 spreads: UN education agency: https://news.un.org/en/story/2020/04/1062232

Rosen, K. (2020, November 12). *attentionfwd*. 2020: Lost year of schooling or year one in a new era of education? https://attentionfwd.com/insights/2020-lost-year-of-schooling-or-year-one-in-a-new-era-of-education/

Sishley, S. (2020, May). *E-Learning Solutions*. Emergency Remote Learning Compared to Online Learning: https://learningsolutionsmag.com/articles/emergency-remote-learning-compared-to-online-learning

Stevens, E. (2020, November 18). *Careerfoundry*. What Is Design Thinking? A Comprehensive Beginner's Guide: https://careerfoundry.com/en/blog/ux-design/what-is-design-thinking-everything-you-need-to-know-to-get-started/

Tschimmel, K., Santos, J., Loyens, D., Jacinto, A., Monteiro, R., & Valença, M. (2018). *D-Think: Design Thinking Applied to Education and Training*. European Commission.

UNESCO. (2020). *COVID-19 Impact on Education*. Education: From disruption to recovery: https://en.unesco.org/covid19/educationresponse

Waloszek, G. (2012). *SAP User Experience Community*. Introduction to Design Thinking: https://experience.sap.com/skillup/introduction-to-design-thinking/

Yang, K. C. (2020). *What Can College Teachers Learn From Students' Experiential Narratives in Hybrid Courses? A Text Mining Method of Longitudinal Data. In Theoretical and Practical Approaches to Innovation in Higher Education.* IGI Global. doi:10.4018/978-1-7998-1662-1.ch006

ADDITIONAL READING

Brown, T. (2008). Design thinking. *Harvard Business Review*, *86*(6), 84. PMID:18605031

Brown, T., & Katz, B. (2011). Change by design. *Journal of Product Innovation Management*, *28*(3), 381–383. doi:10.1111/j.1540-5885.2011.00806.x

Brown, T., & Katz, B. (2019). *Change by design: how design thinking transforms organizations and inspires innovation* (Vol. 20091). HarperBusiness.

Brown, T., & Wyatt, J. (2010). Design thinking for social innovation. *Development Outreach*, *12*(1), 29–43. doi:10.1596/1020-797X_12_1_29

Liedtka, J. (2014). Innovative ways companies are using design thinking. *Strategy and Leadership*, *42*(2), 40–45. doi:10.1108/SL-01-2014-0004

Razzouk, R., & Shute, V. (2012). What is design thinking and why is it important? *Review of Educational Research*, *82*(3), 330–348. doi:10.3102/0034654312457429

Thoring, K., & Müller, R. M. (2011). Understanding design thinking: A process model based on method engineering. In *DS 69: Proceedings of E&PDE 2011, the 13th International Conference on Engineering and Product Design Education, London, UK, 08.-09.09. 2011* (pp. 493-498).

Tschimmel, K. (2012). Design Thinking as an effective Toolkit for Innovation. In *ISPIM Conference Proceedings* (p. 1). The International Society for Professional Innovation Management (ISPIM).

Chapter 3
Problem-Based Learning (PBL): A Deep Approach to Learning in the 21st Century

Mahmoud Hawamdeh
https://orcid.org/0000-0001-5321-2400
Al-Qud Open University, Palestine

Idris Adamu
Aminu Saleh College of Education, Nigeria

ABSTRACT

This chapter discuss how Problem-Based learning (PBL) helps to achieve this century's approach to teaching and learning for students in higher educational institutions. If adopted, this method of teaching will enable student to attain learning skills (skills, abilities, problem solving, and learning dispositions that have been identified) to acquire a lifelong habit of approaching problems with initiative and diligence and a drive to acquire the knowledge and skills needed for an effective resolution. And they will develop a systematic approach to solving real-life problems using higher-order skills.

INTRODUCTION

Problem-based learning (PBL) is a type of education that drives everyone to learn through a real problem. In PBL, the window instead goes out, you learn the basics for a meaningful, but the unstructured procedure for solving problems Initially, PBL has been developed from the educational need to allow medical students to acquire a knowledge of the fundamental sciences while helping to improve their clinical abilities

DOI: 10.4018/978-1-7998-4534-8.ch003

that would be more lasting concurrently. Although a particular need has been met, PBL has the foundation of sound ideas and paradigms in education. This chapter is going to be Assist the reader in recognizing PBL's theoretical foundations; we are starting to understand that PBL can function, and analyzing the educational system The idea it is founded on will help us to understand why as well as why. Allow us to diagnose and enhance PBL apps when things are not pretty much going as expected.

In education, the important problem was the evaluation of 21st-century skills (Geisinger. 2016). Since it was a central term and slogan in the field of 21st-century skills dependent on education (Greiff &. Kyllonen, 2016: Teo (2019) also noted that teachers have identified strategies over the past two decades to educate students and prospective teachers on how to manage the increasingly globalized environment and the 21st century-related inter-connected landscape. Since students need to have these abilities in the future (Larson & Miller, 2011). Besides, the Wang et-al (2018) announced that skills in the education of the 21st century have also become an important topic around the world. Thus these explanations indicate that other 21st-century skills studies need to be carried out from different backgrounds and important areas, including 21st-century skills studies involving prospective teacher students in the field of learning (A. Sujana*, A. Permanasari, W. Sopandi, 2012).

The teaching and learning methods of the twenty-first century have progressed in such a way that many writers consider them to be a new way of accessing knowledge and as a result, transforming learning. Educational activities, both managed (presential, semi-presential, and online teaching) and non-regulated (Massive Open Online Courses MOOCs), are usually required either completely or partially through interactive platforms known as learning management systems (LMS) such as Moodle (Allen and Taylor, 2017).

This reality meant a change in the way knowledge is transmitted and processed. Due, where necessary, to the educational characteristics of higher education, LMS is more critical and needs changes in the teaching and learning styles of most teachers and learners (Sáiz-Manzanares, 2018).

Globally, students learn by debating professionally related topics that facilitate the application and incorporation of information, which is supposed to inspire students to pursue an in-depth Learning is designed in which learners are intrinsically interested and attempt to understand what is being studied. Higher institutions today face the challenge of educating students to be lifelong learners and specialists in their field Promoting and encouraging the growth of Learning and development skills, such as problem-solving and critical thinking, have become a core focus of higher education in the 21st century.

Asikainen, (2014). a good approach to learning and advanced education should include learners in deep learning. Problem-based learning (PBL) is essentially a learning framework plan that incorporates a few insightful procedures to maximize student-centered learning outcomes, rather than just gathering knowledge. According to Serife (2011). Problem-based learning has a positive effect on the adoption of a primarily deep learning approach by learners and has a detrimental effect on the adoption of a surface approach to learning.

Dinsmore and Alexander (2012) suggest that one field requires a crucial discussion of deep and surface learning if exploring the classroom instruction will have any impact on practice.

From their analysis, Dinsmore and Alexander (2012) stressed It is important for research to clearly define what deep learning means, to study deep learning within a particular learning context, as deep learning can be influenced by the context of the learning environment, and to test deep learning through valid instruments. This chapter, therefore, discusses the possibility of using problem-based learning (PBL) as an in-depth method to 21st-century skills among learners. in higher education institutions.

BACKGROUND

The notion of Deep and surface learning varies through investigation; as does how these concepts are tested. How we prepare youthful learners for life, career, and work every day is a question that reacts to the growing world of the 21st century, workers, authorities, caregivers, academics, and college students even are asking. There is also insufficient confirmation of the strength of the methods used to test deep learning. Another reason is that the context in which the studies are performed varies frequently, since deep learning may vary across settings and academic fields.

According to (Lonka and Lindblom-Ylanne 1996, Biggs et al. 2001; Entwistle and McCune 2004; Loyens et al. 2013), a deep learning method has now been defined as the purpose of a student to comprehend content along with the processes of connecting and structuring ideas, searching for underlying values, evaluating relevant facts, and assessing information critically. Whether students handle their learning is seen as variable and affected by learning environments, the perspective of learners on these components, and the characteristics of students, problem-solving skills, lifelong learning, and technology learning.

PROBLEM-BASED LEARNING (PBL)

When questioned to provide a summary of problem-based learning on the introductory query, We immediately agreed with this newspaper, thinking it was a perfect opportunity to write about a newspaper. I'm interested in the subject. When did we start worrying about What's PBL? "Some of the items we learned about the PBL were taught via PBL, but reading a lot of papers, so much more was learned from reading experts with decades of experience in the field of study, and problematic practice. "It became apparent that we had a problem.

These writers have also begun their papers with a context-setting debate, the authors also began their works with a contextual discussion.

What else do you say about him? What is PBL? Since the 1970s, PBL has been successfully implemented in many regions. Theory and implementation of PBL cannot be viewed unanimously. The theoretical foundation of PBL was not evident in early PBL literature (Rideout & Carpio 2001). Skelly, the pioneer of PBL, states that he has no experience in cognitive science or educational psychology like other developers of the original McMaster PBL software

They just thought that the medical education of their students by using clinical questions would become more meaningful and relevant for small groups (Barrows 2000). PBL has traditionally been profoundly affecting cognitive psychology in medical education (Norman & Schmidt 1992; Schmidt 1983; Schmidt 1993). Much presently with PBL spreading to further fields, other educational theorists, such as Dewey (1938) and involvement stressed different aspects of teaching and learning, extracted theoretical reasoning from other theories (Newman, 2003).

Related learning problem (PBL), during the period of John Dewey, has been recognized. PBL is the presentation leaners have genuine and concrete problems that promote analysis and study. Similarly, a study conducted by Gorghiu et al (2015). the view that PBL is defined as called "research-based learning," an efficient way for learner's s to work in a variety of fields, or curriculums, that can improve fundamental skills. This claim is backed by the PBL, Arend (2008b), which is a learning model that enables students to build their expertise, greater abilities, and study, develop learners independence to build students' self-assurance in real and practical problems. PBL is solemnly a pioneering learning model for active and imaginative learners in 2014, according to Saragih and Habeahan (2014).

Abdullah et al. (2010) also indicated the potential of PBL to improve collaboration in addressing communication issues and skills. PBL would encourage learners to build their problem-based skills and actively engage in the development of a job or product after learning. (Siagian et al., 2019).

As the Real-life problems arise in each area, these sort of problems has to be sold upon graduation. PBL has proven popular in medical, electronic engineering, science, and the education of art and humanities. It should be observed that these are knowledge, process-oriented forms of education where language becomes a tool if we restrict our study to higher education in foreign languages or in the ESP. Problem-based instruction is a student-centered pedagogy in which students learn from the subject by solving an open problem in prompts., (Wikipedia 2020).

Problem-based learning (PBL) is a learning methodology that is used in a variety of universities around the world. The topic is addressed initially before beginning a study since previous student experience does not solve the problem. P.B.L. set by the student to discuss it professionally.

(Merrill 2012; Van Merrienboer and Kirschner 2013), the method of PBL is consistent with existing approaches to instructional design, demonstrating how important it is to fully understand the whole issue so that information and behaviors are avoided and incorporated.

In other words, PBL would urge a deep-rooted approach to learning because students explore links between ideas and standards, use distinctive literature materials, apply these ideas and standards in subjects being addressed and organize knowledge and skills. As a time of deep change in the Rich Environments for learning requirements, revolutionary multidimensional learning needs to be explored. The multidimensional dimensions of learning can be viewed in Figure 1. Brief overview of the main concerns associated with the various aspects will follow.

Assume that educators have taken seriously the often-mentioned goal, that is, that they want to really help their students to solve problems, armed with practical skills that can be enabled to help them ensure and identify issues, build work hypotheses, collect, analyze data, synthesize findings into solutions and assess or explain them.

GOAL OF PROBLEM-BASED LEARNING

Problem-based curricula offer driven experience to students learning by solving complex problems in the real world. PBL has been developed for many important objectives (Barrows and Kelson, 1995). It's meant to assist learners on the following goals:

1. To develop a wide and versatile base of knowledge.
2. Develop successful troubleshooting capabilities.
3. Develop lifetime skills for self-direct learning.
4. Become successful employees.
5. Get inspired to learn intrinsically.

Building a wide and versatile knowledge goes beyond knowing the details of a specific area. The integration of knowledge across multiple domains is important. The deeper concepts in a field are regularly organized into this awareness (Chi et al., 1981)(Cindy E. Hmelo-Silver1, 2004). Common sense indicates that learners should be encouraged to develop versatile learning. We must integrate learning into information and productive problem-solving skills. Contexts that enable these abilities to be used. Experiments in the laboratory have It has shown that this is indeed the case.

Figure 1. Multidimensional Aspect of Learning for PBL for the 21ST Century.

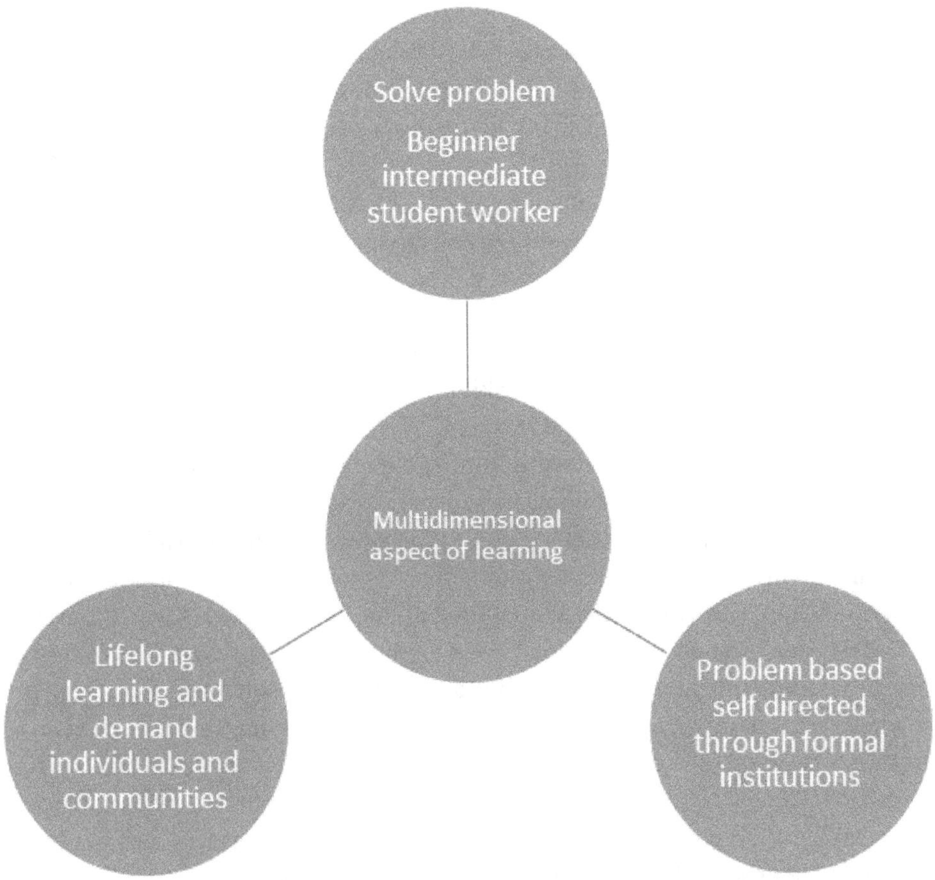

CHARACTERISTICS OF PROBLEM-BASED LEARNING

For more than a period, we have worked with teachers in post-primary school history on a changing investigation line this refers to research-based ideas for the plan and trying of a problematic learning environment and a variety of realistic forms of teaching proficiently to incorporate a new model of teaching based on problems.

Historical investigation (Brush & Saye, 2014) PBL is a learner-centered educational (and curricular) Approach that enables learners to evaluate, mix Theory and practice, and use expertise and skills to improve a viable solution to a given problem. A teacher who directs the learning process is crucial to the success of the learning process method and gathers malicious (often interdisciplinary) problems.

At the end of the research experience, comprehensive debriefing is performed and carried out.

The features and characteristics needed for a good PBL have been identified by several writers.

The instruction approach. In short, the Readers are advised to read sources Quotes do not comply with the degree of information that the writers offer. Boud's and Feletti's (1997) presented a list of methods deemed to be characteristic of philosophy, approaches,

Problem-based learning tactics and tactics. The methods defined by Duch, Groh, and Allen (2001) used in the field of PBL and the special skills developed, including critical thinking capacity, analysis to discover, analyze, and use effective learning to solve complex real-world problems,

Resources; function together, show strong teamwork and communication skills demonstrate effective communication skills; Using the mastery of the material and to improve intellectual skills to the learners.

Sage & Torp (2002) PBL has been described as concentrated, experience-oriented learning to investigate and overcome messy real-world problems. PBL was defined by Hmelo-Silver (2004) as a teaching process, in which students learn by facilitated problem-solving centers a complex problem which does not have a clear correct answer, and describes the student as a problem that affects Solvers who are trying to find the root problem and the conditions for an effective solution. She observed that learners working in collaborative groups decide how to solve a dilemma, engage in self-directed training, and use their new skills. Understanding and concentrate on the question. What they discovered and the usefulness of the methods employed.

A list of Common PBL Basics, reduced to bullet points (http:/www.pbli.org/pbl/genericpbl.htm) to provide supplementary information and services, each of these important features has been expanded briefly.

The following are from a learning climate for PBL:

1. **Problem-focused:** students started learning through dealing with simulations of an authentic, malformed problem. Curriculum organization and learning skills Around questions Rather than a hierarchical list of concerns. Awareness is then learned in the context of the problem, and understanding and problem are interlinked. Sensitivity is inspired by the problem and the problem has been remedied.
2. **Student-centered:** Learning is not determined by faculty Operations, but function in a supporting capacity instead.

3. **Self-directed:** Learners are responsible for the development of learning challenges and processes internally and collaboratively over self-evaluation and peer evaluation and accessing their own understanding and learning materials through experience rarely are appropriate assignments given.
4. **Self-reflective:** students watch their comprehension of study change learning techniques.
5. **Facilitative:** The facilitators are teachers facilitating supporting and modeling processes of reasoning Interpersonal dynamics and sample structures.

SUITABILITY FOR PROBLEM-BASED LEARNING

Problem-based learning as an educational approach which, similar to teaching pedagogy, is not uniformly applicable to various problems of learning. PBL's primary objectives are to boost the Application of information, Problem-solving, and acquiring skills that are self-directed by students through Necessitating them to articulate, comprehend, and address issues actively. The issue with PBL is based, where learners begin learning by resolving some authentic issue simulations. Instead of being a hierarchical list of topics, there is a mutual relation between the knowledge and the question, the subject matter material, and the skills to be mastered are structured around issues. Learning is motivated by the problem and has been applied to the problem. PBL is often student-centered, requiring students to direct their own learning. Deciding what they're doing and they don't know about the subject (Brush & Saye, 2014), It is possible to summarize the basic principles s follows issues with PBL should be Open-ended, badly organized still with a mild level of structurally; The degree of complexity should, however, be complex. Be challenging and inspiring, interacting with the desires of students. Provide opportunities for students from different perspectives or disciplines to discuss the issue; tailored to the prior experience of students; tailored to the cognitive growth and preparation of students; Authentic contextualized as to the future or possible workplace for student's workplaces.

DEEP LEARNING

The term 'deep learning' has become widely accepted as a core concept in the growth and user experience of the learning process. In response to the changing world of the 21st century, how we prepare learners for life, leisure, and work today is a question that employers, governments, parents, educators and students themselves pose. With changes in teaching, learning, and evaluation, the education curriculum is moving quickly to meet the needs of students. The changing role of the student, who is never again a passive recipient of learning, but a functional part of an aspect of the transformation phase, from plan to execution, is necessary.

Deep learning is a part of a large family of learning machines focused on artificial neural networks with representational learning this is also known as deeply organized learning. Learning can be tracked, semi-monitored, or unmonitored.

In the areas of computer view, machine vision, speech recognition, natural speech recognition, audio recognition, social network filtering, translation machine, bioinformatics, drug design, medical image analysis, content inspection, the implementation of profoundly learning architectures such as deep neural networks, profound belief networks, recurrent neural networks and coagulation of neural networks (Wikipedia 2020).

The area of artificial intelligence is primarily when computers can perform tasks that are usually humane. It involves machine learning, in which machines can learn from experience and develop skills without involving humans. Deep learning is a sub-set of machine learning in which vast quantities of data are used to train artificial neural networks, inspired by the human brain. Like how we learn from experience, each time we tweak the deep learning algorithm to boost the performance, it is a process that is repeatedly performed. We are talking about proper learning,' because there are different deep) layers in neural networks that allow learning. Any problem that needs "thinking" to be solved is a problem that can be tackled by deep learning (Lucia et al., 2021).

Deeper learning can ensure that students have transferable information or the ability to use their understanding and skills to solve challenges and manage new situations. As an ability set of the 21st century, it should be a central aspect of public education learning experience.

Deep learners are thought to be those who approach information and learning by relating new knowledge to previous knowledge. This is expressed by the transformation of knowledge by Entwistle (2000). Deep learners also connect theoretical concepts to everyday experience; they differentiate between evidence and argument; they organize and arrange information into a coherent whole; they intertwine knowledge from different sources; and they are self-motivated (Atherton 2005). These qualities are highly desirable, as represented by a flexible and independent learner who will excellence in a changing society.

To evaluate the possible benefits for the learner and for the broader community, a better understanding of deep learning is needed. While no sole definition about deep learning based on the mentioned criterion. According to Simms (2006) "Deep learning is secured when, through personalisation, the conditions for student learning are transformed."

It was helpful because it shows the value of deep learning conditions and their close relationship to personalisation. The emergence of the word personalisation reflects a move towards a form of education that is much more open-minded and inclusive. Focus on the user described in deep learning makes this a potential source of personalization.

Deeper learning. As previously noted, deeper learning competencies include Academic content mastery, strategic thought and problem solving, productive communication, expertise in teamwork, learning how to read, and academic ways of thinking (William and Flora Hewlett Foundation, 2013). The National Registration Council (2012) extends this definition of deeper learning "as the process through which an individual becomes capable of taking what is learned in one situation and applying it to new situations (i.e., transfer)" (p. 5). Psychological, intellectual, and interpersonal aspects of information and skills are best integrated in combination in its seminal Education in 21st century life and work.

In a detailed report examining these competencies across high schools committed to deeper learning, the American Institute for Research (AIR) does not cite access to technology as requisite to achieving these competencies (Huberman, Bitter, Anthony, & O'Day, 2014). Instead, they find that schools that ensure deeper learning embed opportunities for project-based learning, emphasize mastery of core academic content, focus on interpersonal and intrapersonal skill development, target learning to learn strategies, develop academic mindsets, and foster cultures that value and support student-centered instructional strategies (Huberman et al., 2014).

Per the OECD (2013), shifts in policy and practice that focus on 21st-century skills in education can lead to growth in student achievement. The OECD (2013) emphasizes that educators need to build pedagogies that enable learners to become successful problem-solvers, risk-takers, partners, problem-solvers, and "knowledge workers." (p. 35). Building upon these 21st-century skills, the William and Flora Hewlett Foundation (2013), emphasize The need to foster important core competencies such as academic teaching methodologies, Critical thinking and problem-solving, strong leadership skills in teamwork, understanding how to read, and mind-set comprehension skills.

Deep learning by coevolutionary by learning from raw data, Vision was revolutionized by neural networks (deep ConvNets). The use of deep ConvNets for end-to-end EEG analysis is important, but there is still a need for a better understanding of how to design and train ConvNets for end-to-end EEG decoding and how to visualize the insightful EEG features that ConvNets learn. (Schirrmeister et al., 2017).

Collectively, the Hewlett Foundation (2013) and their supporters refer to these competencies as deeper learning. Arguably, the increase of one-to-one devices can support a transformation of teaching and learning when the focus is on 21st-century skill acquisition and the fostering of deeper learning. However, mission-driven technology integration is currently scattered at best (Friday Institute, 2011; Groff, 2013; Harris, Mishra & Koehler, 2009; Mishra et al., 2011; Topper & Lancaster, 2013).

PROBLEM-BASED LEARNING AS AN INSTRUCTIONAL DESIGN MODEL

The principles of education design used in the context of Lebow (1993) values may contribute to a broad range of problematic learning environments (PBL) as an overall model in medical education developed in the mid-1950s, and refined by more than 60 medical schools. since that time. The most common use of the PBL strategy has been developed

By dividing the problem-based learning course into six stages, teachers will begin to plan, conduct, and analyze problem-based learning in their course as follows:

Phase One: Classify Results/Evaluation

Problem-based learning paroxysms better with sequence results that are process-oriented, like teamwork, study, and solve problems. Therefore, assist learners in learning content, definition, or disciplinary behaviors such as writing or communication. You will create process and end assessments for the measurement of student learning after determination of whether your course has results that match PBL. Possible PBL evaluations are group contracts, self-assessment forms, learning reflection, sampling, and rubrics.

Phase Two: Designing the Scenario

Next, with jointed combined issues that will arise from student thinking, the problem-based learning scenario will be designed. Think of a true, complicated problem relevant to the content of the course. Identifying many issues in our fields is rarely difficult; the trick is to For the learners, write a hypothetical that will evoke the kinds of thinking, interaction, research, and learning that must take place to meet the outcomes of learning. Scenarios should be inspiring, informative, and good for debate should be created.

Phase Three: Introduction to the Problem –Based Learning

If your learners find Problem-based Learning new, the teacher should practice in the dining hall with an "easy problem" such as an explanation of long lines. Introduce the assignment descriptions, headings, and timelines for grouping learners and allowing time to engage in the abbreviation of Problem-based Learning. Let the groups read the scenario, (s). One scenario may be created and each group discussed in its own right or several scenarios structured to resolve a single topic for every group to explore and investigate.

Phase Four: Problem-based learning research

Problem-based learning study starts with brainstorming sessions in small groups where students can ascertain the problem and what the problem is (background knowledge), what the study topics ought to learn better, and where data is needed to find (databases, interviews, etc.). Groups should be writing a report or a research question about the issue. You would need support. Consider your study: the procedure can be unguided or far too detailed without sufficient research questions. Students should determine community roles to learn the topics required for a complete comprehension of their complications. Learners then establish an original "test" theory as they are searching for a solution. Recall: questions and theories will shift when students discover evidence that disconfirms their initial convictions.

Phase Five: Performance of the Products

After the study, learners produce goods and presentations that replicate their research, solutions, and learning. It is entirely up to you to develop the summative assessment. This transfer is approached as a fair trial. Students find tools to gain context information that supports their interpretation and then present their results as study posters to the class collaboratively, providing one or two feasible solutions.

Phase Six: Assessment

Assess the products and outcomes of the groups during the problem-based learning evaluation phase. Using rubrics to assess whether the problem, context, research methodology, ideas (feasible and research-based), and tools have been clearly articulated by students and to decide whether all community participants have participated meaningfully. Every day, and at the end of the process, you should consider allowing your students to focus on their research (including what they have learned about the material and the study process).

While problem-based learning is described as phases, it ultimately operates cyclically. For Eg You could take a course in economics and develop a scenario for busy campus pathways. Following the reading of the scenario, the groups generate temporary ideas about why the sidewalks are crowded and how to solve the problem. If a group says it is crowded because the streets are too narrow and the remedy spreads resulting in economic and environmental impact analysis will tell them that it is not practical to expand the sidewalks. They should jump back to stage four, explore a new theory, and launch a different course of study.

HOW PROBLEM-BASED LEARNING PBL HELP TO ACHIEVE THE 21ST CENTURY

To help them, diverse types of problems need diverse types of learning methods and multiple social and engineering settings. Apart from the classroom, a lot of learning and problem solving takes place because people examine meaningful issues individually and participate in group activities by using media and technology extensively. Many prior educational frameworks have been founded on the premise problem-based learning is central to teaching. (Thomas & Brown, 2011).

Science technology engineering and mathematics STEM use problem-based learning It is difficult to distinguish how STEM affects the improvement of students in these fields, which has been shown to increase 21st-century skills. Participating in classroom activities helps students through PBL to develop skills of the 21st century in a meaningful way through deep learning. As Loveland & Dunn (2014) claims, STEM education also focuses on integrating them to solve real-world issues. Engineering habits are also integrated into design projects in all classrooms. Such habits of mind have also integrated skills such as imagination, teamwork, and communication from the 21st century that engineers use daily.

To make intelligent choices, analytical thinking and problem-solving are knowledge that evaluates and critique circumstances. Rational thought and problem-solving are the behaviors of the mind, such as continuing when tasks are challenging, controlling impulsivity, intellectual flexibility, according to Costa & Kallick (2007). which are the 21st-century skills achieved through the deep learning approach.

According to Kelly, (2014), PBL increases technological literacy which is crucial to help students developing critical thinking, which is Teaching strategic thinking to address issues related to technology and technical processes. PBL does not only need more critical thought but to solve the problem, it also needs greater levels of imagination. Another ability that can be developed by an acquaintance to PBL generated by STEM is innovation, which includes developing fresh and unique ideas and products. Collaboration and communication skills are also developed through these practices. For communicating with others and exchanging ideas efficiently, these skills are important.

While learners often believe it's harder to work in groups than to work on their own, students must develop good team skills to work efficiently and effectively with others (Morrison et al., 2015). A 21st-century student can research autonomously, but also in groups, well (Prettyman et al., 2012). Therefore, students need to have opportunities to bring their skills and know-how to group environments and to share these skills to achieve a shared group goal.

Jensen et al. (2014) found that when preparing for assignment essays students were more inclined than a multiple-chosen assessment to use a deep approach that they found to evaluate higher cognitive levels of processing.

Moreover, PBL improves deep knowledge in most studies and does not affect learning on the surface, this study also demonstrates, as Dinsmore and Alexander conclude, that studies sometimes lead to uncertain and conflicting results (2012). One explanation is that out of 21 studies, only three studies commented on data validity and eight on data reliability. As previously concluded by Dinsmore and Alexander, evidence of validity was sometimes missing (2012).

Within this study, we discussed deep learning, being PBL, in a particular sense. Although the studies have shown that the PBL is applied differently over a variety of studies, although the research included only studies that fulfilled our PBL definition. The findings of the study have shown that PBL is being implemented in different studies. Furthermore, it was argued in one study that students already had high deep learning scores, which may find it harder to further enhance deep learning (Reid et al. 2005).

CONCLUSION

This Chapter presently aimed to show why PBL is a comprehensive educational model and to explain how PBL is compatible with teaching concepts that derive from constructivism. We were trying to establish a direct relationship between theory and reality. The features of the PBL environment include the continuous participation of students in projects and events that are authentic for their environments. The emphasis is on learners as builders of their own knowledge in a way like that in which they apply the knowledge. Students are encouraged to think and track their own intelligence, i.e. function objectively and creatively on a metacognitive stage. Social background negotiations are an integral component of the problem-solving team organization and the facts of the case are facts only if they are decided by the group. PBL, as we have described it, compared to a variety of other problems or case-based approaches.

Problematic problems play a part in the success of the learning results of learners in all types of teaching approaches that utilize issues. An issue with a suitable differential culture The level of complexity will surpass the readiness of the learners and lead to failure, while an insufficient level is within the cognitive readiness of the learners and can therefore be solved. The problem complexity assessment aims to help researchers identify the types of problems in PBL settings, which are most effectively used. The technique's difficult assessment of the issue serves as a reference for designers and educators to chart the degree and the norm.

Promoting the intended educational purpose and function of the issue and maintaining an appropriate degree of difficulty for the intended level of the course, to balance the essence of the subject area with the nature of the complexity of a given problem. We intended to examine the architectural problem of the problem of complexity, which in the layout of courses with PBL has not been fully utilized.

Individuals differ, specifically concerning learning, with varied demands and needs. In support of a rich landscape, New media provide the basis for social and technological environments of learning.

These explanations of PBL's characteristics clearly describe 1) the position of the tutor as a tutor.

A learning facilitator, 2) the learners' roles in their learning to be autonomous and self-regulated, and 3) the basic Components in the production of ill-structured learning training issues as the motivating force for investigation. For several teachers, the challenge When they follow a PBL strategy, the change from instructor to information is to make Supplier to mentor as supervisor and learning facilitator (Ertmer & Simons, 2006: (Savery, 2006)). If PBL teaching was as easy as posing a "problem" to the learners and students.

It can consistently be used to track awareness and self-evaluation with a high level of self-monitoring. The self-regulation of several teachers would then require early withdrawal. That's the truth to help new PBL students, large educational facilities are important to facilitate the self-reliance of young people, the development of problem-solving skills, self-managing learning skills, and skills in teamwork and cooperation. In recognition of their vital role in fostering the experience of PBL learning, TEACHING Institutions pursuing a curriculum and instruction strategy for PBL (including the above) have established robust tutoring programs.

The resource is Barrow's (1988) (Savery, 2006) tutorial process, which explains the importance of A mentor as a metacognitive teacher for the learner. Although the change of education patterns occurs at an ice-cream pace, it takes time for organizations to participate completely in problem-based learning.

There are conversely, some narrowly related teaching methods focused on learners, such as

Project-based learning that is used in a project, case-based learning, and inquiry-based learning

A selection of content domains that can start moving learners along the road to becoming

In their learning, the more self-directed. I will discuss some similarities and similarities in the next segment. Differences between such methods(Savery, 2006).

It is well understood that well-developed 21st-century knowledge is the secret to success in a technical and global society. However, due to the methods of teaching, not all students had learned these core skills. More students receive career advice through PBL, STEM, including digital literacy, critical thinking, problem-solving, innovation, and 21st Century cooperation.

Along with the challenges of the 21st century, higher education institutions are required to include in their curricula problems solving preparation, teamwork, and decision making. PBL has proven to be a highly effective method, as it can turn information that is obtained passively into active information. The types of activities used by PBL help improve and train skills, including efficient communication, effective presentation, and good meeting skills, while also preparing people for teamwork. Furthermore, this form of study prepares students for personal lifelong learning.

Finally, with the implementation of PBL in various classes, students have become more optimal, more independent, and experienced, who can withstand their opinions and develop so valuable abilities that increase motivation and productivity for learning and thus achieve their skill and excellence.

The basic processes in PBL theoretically support the development of critical thinking by students according to the design applied,

Empirical evidence is usually inconclusive in the interpretation of the consequence. of PBL on the critical thinking capacity of students specifically studies out of the healthcare sector,

More indication gives that PBL needs long-term exposure to promote the critical thinking of students It means that in the course of PBL, the essential elements contributing to PBL efficiency must be carefully developed and discussed. This includes the functions of the facilitator in the media literacy of the students particularly in the meta-cognitive thinking of the students. With all of these considered carefully, PBL will succeed as theoretically noted (Yamin & Masek, 2011).

REFERENCES

Allen, L. E., & Taylor, D. M. (2017). The role of the academic library information specialist (LIS) in teaching and learning in the 21st century. *Information Discovery and Delivery*, *45*(1), 1–9. doi:10.1108/IDD-09-2016-0030

Brush, T., & Saye, J. (2014). An Instructional Model to Support Problem-Based Historical Inquiry: *The Persistent Issues in History Network. The Interdisciplinary Journal of Problem-Based Learning*, *8*(1), 3–7. doi:10.7771/1541-5015.1409

Christiansen, E. T., Kuure, L., Mørch, A., & Lindström, B. (2013). Problem Based Learning for The 21st Century New Practices and Learning Environments. Aalborg University Press.

Costa, A., & Kallick, B. (2007). *Describing 16 habits of mind*. Retrieved November 13,2020, from http://schoolnet.org.za/teach10/resources/eo/resources/thinking/Habits_of_Mind.pdf

Dinsmore, D. L., & Alexander, P. A. (2012). A critical discussion of deep and surface processing: What it means, how it is measured, the role of context, and model specification. *Educational Psychology Review*, *24*(4), 499–567. doi:10.100710648-012-9198-7

Entwistle, N., & Peterson, E. (2004). Conceptions of learning and knowledge in higher education: Relationships with study behavior and influences of learning environments. *International Journal of Educational Research*, *41*(6), 407–428. doi:10.1016/j.ijer.2005.08.009

Groff, J. (2013). *Technology-rich innovative learning environments*. Paris, France: OECD Publishing. Retrieved November 23, 2020, from http://www.oecd.org/edu/ceri/TechnologyRich%20Innovative%20Learning%20Environments%20by%20Jennifer%20Groff.pdf

Harris, J., Mishra, P., & Koehler, M. (2009). Teachers' technological pedagogical content knowledge and learning activity types: Curriculum-based technology integration reframed. *Journal of Research on Technology in Education*, *41*(4), 393–416. doi:10.1080/15391523.2009.10782536

Hmelo-Silver, C. E. (2004). Problem-Based Learning: What and How Do Students Learn? *Educational Psychology Review*, *16*(3), 235–266. doi:10.1023/B:EDPR.0000034022.16470.f3

Jensen, J. L., McDaniel, M. A., Woodard, S. M., & Kummer, T. A. (2014). Teaching to the tes or testing to teach: Exams requiring higher order thinking skills encourage greater conceptual understanding. *Educational Psychology Review*, *26*(2), 307–329. doi:10.100710648-013-9248-9

Loveland, T., & Dunn, D. (2014, May/June). Teaching engineering habits of mind in technology education. *Technology and Engineering Teacher*, *73*(8), 13–19.

Lucia, S., Navarro, D., Karg, B., Sarnago, H., & Lucia, O. (2021). Deep Learning-Based Model Predictive Control for Resonant Power Converters. *IEEE Transactions on Industrial Informatics*, *17*(1), 409–420. doi:10.1109/TII.2020.2969729

Marra, R., Jonassen, D., Palmer, B., & Luft, S. (2014). Why Problem-Based Learning Works: Theoretical Foundations. *Journal on Excellence in College Teaching*, *25*, 221–238.

Merrill, M. D. (2012). First Principles of Instruction: Identifying and Designing Effective, Efficient, and Engaging Instruction. Hoboken, NJ: Pfeiffer (John Wiley & Sons).

Mishra, P., Koehler, M. J., & Henriksen, D. (2011). The seven trans-disciplinary habits of mind: Extending the TPACK framework towards 21st century learning. *Educational Technology*, 22–28.

Morrison, J., Roth McDuffie, A., & French, B. (2015, May). Identifying key components of teaching and learning in a STEM school. *School Science and Mathematics*, *115*(5), 244–255. doi:10.1111sm.12126

Newman, M. (2003). *A pilot systematic review and meta-analysis on the effectiveness of problem-based learning*. Retrieved November 10, 2020 from https://www.heacademy.ac.uk/assets/documents/subjects/medev/pbl_report.pdf

OECD. (2013). *PISA results: Ready to learn: Students' engagement, drive and Self-beliefs* (Vol. 3). OECD Publishing. doi:10.1787/9789264201170-

Prettyman, S. S., Ward, C. L., Jauk, D., & Awad, G. (2012, Fall). 21st century learners: Voices of students in a one-to-one STEM environment. *Journal of Applied Learning Technology*, *2*(4), 6–15.

Putnam, A. R. (2011). Problem-Based Teaching and Learning in Technology Education. *5th Slovakian-Hungarian Joint Symposium on Applied Machine Intelligence and Informatics*.

Reid, W. A., Duvall, E., & Evans, P. (2005). Can we influence medical students' approaches to learning? *Medical Teacher*, *27*(5), 401–407. doi:10.1080/01421590500136410 PMID:16147792

Sáiz-Manzanares, M. C. (2018). *E-Project Based Learning en Terapia Ocupacional: una Aplication en la Asignatura "Estimulacion Temprana*. Servicio de Publicaciones de la Universidad de Burgos.

Savery, J. (2006). Overview of Problem-based Learning: Devinition and Distinction Interdisciplinary. *Journal Problem-Based Learning*, *1*(1), 9–20. doi:10.7771/1541-5015.1002

Schirrmeister, R. T., Springenberg, J. T., Fiederer, L. D. J., Glasstetter, M., Eggensperger, K., Tangermann, M., Hutter, F., Burgard, W., & Ball, T. (2017). Deep learning with convolutional neural networks for EEG decoding and visualization. *Human Brain Mapping*, *38*(11), 5391–5420. doi:10.1002/hbm.23730 PMID:28782865

Siagian, M. V., Saragih, S., & Sinaga, B. (2019). Development of Learning Materials Oriented on Problem-Based Learning Model to Improve Students' Mathematical Problem Solving Ability and Metacognition Ability. *International Electronic Journal of Mathematics Education*, *14*(2), 331–340. doi:10.29333/iejme/5717

Sujana, A., Permanasari, W., & Sopandi, A. M. (2012). Article. *Jurnal Pendidikan IPA Indonesia, 1*(1), 91–97. Retrieved November 24, 2020, from: http://journal.unnes.ac.id/nju/index.php/jpii%0ALITERASI

Thomas, D., & Brown, J. S. (2011). *A new culture of learning: Cultivating the imagination for a world of constant change. Lexington*. CreateSpace.

Topper, A., & Lancaster, S. (2013). Common challenges and experiences of school districts that are implementing one-to-one computing initiatives. *Computers in the Schools*, *30*(4), 346–358. doi:10.1080/07380569.2013.844640

Van Merrienboer, J. J. G., & Kirschner, P. A. (2013). *Ten steps to complex learning (2nd Reviewed.)*. Routledge.

Yamin, S., & Masek, A. (2011). The Effect of Problem Based Learning on Critical Thinking Ability: A Theoretical and Empirical Review. *International Review of Social Sciences and Humanities, 2*(1), 215–221. Retrieved November 01, 2020 form www.irssh.com

Chapter 4
Contributions of Information Design to Popular Scientific Communication:
A Semiotic Approach

Juliana Rocha Franco
State University of Minas Gerais (UEMG), Brazil

Izabela Silva Pinho
State University of Minas Gerais (UEMG), Brazil

ABSTRACT

In a world increasingly saturated with data, scientific communication's success in reaching its target audience is linked to its ability to organize and represent information. Through semiotics concepts, this chapter explores the possibilities of information design as a set of tools for planning the exhibition of scientific content effectively. Information design is an area of knowledge that focuses on the human component's concern and the ergonomic aspects of visual communication. The chapter brings two case studies on infographics and on how information design is applied in popular scientific communication. The research will show that information design, associated with the precepts of plain language, can offer substantial contributions to the work of those who intend to disseminate science to non-specialized audiences.

DOI: 10.4018/978-1-7998-4534-8.ch004

INTRODUCTION

The purpose of the chapter is to present some contributions of Information Design to Popular Scientific Communication. Information Design is an area of Graphic Design that aims to balance, with its target audience, the aspects that involve information systems through the contextualization, planning and production of graphical interface of information. As an area of knowledge, Information Design encompasses the organization and communication of knowledge itself. Its strategies and methodologies can make substantial contributions to Science Communication in its communicative aspect.

Images, graphs, and diagrams have long been used to help communicate various aspects of science and make it accessible to wider audiences through visual language (Estrada & Davis, 2015). Visuals have been present in scientific journal articles since that form of communication emerged in the 17th century (Ostergren, 2013). Modern science has placed images at the center of its communicative processes (Bucchi & Saracino, 2016).

However, methods of communicating scientific information have changed over time. Nowadays, the importance of knowledge organization, visualization, and communication has grown with the advent of new information technologies. The need to structure and communicate information, complex data and systems, a constant challenge of scientific disseminators, has increased significantly in recent years (Gross, Harmon & Reidy, 2002).

Bucchi & Saracino (2016) state that both experts and diverse audiences currently live immersed in a visually dense environment, especially when it comes to science and technology content. The efficiency and accessibility of scientific dissemination today permeates the need to organize this profusion of data that flooded the world in recent decades. Being able to orientate oneself in this ocean of data and information is directly linked to the way they are structured and presented. Moreover, such capacity is indispensable for those who intend to follow the path of continuing education, increasing the possibilities of self-education and informal education, fundamental to lifelong learners.

With that in mind, the chapter will present Information Design procedures that can help to plan scientific communication aligned with the concept of plain language. The goal is to demonstrate that the structuring and presentation of content are as important as the content itself. The scientific communication structured by Information Design can help to promote an inclusive practice of information dissemination that goes beyond the pursuit of the availability of pure and direct information, as well as to seek that individuals can understand, evaluate and use it.

THE IMPORTANCE OF CLARITY AND THE CONNECTIONS BETWEEN INFORMATION DESIGN AND PLAIN LANGUAGE

Several authors (Brockington & Mesquita, 2016; Fischhoff & Scheufele, 2013) define scientific communication as the transmission of scientific information to an unskilled audience. In order to do so, it is essential to re-code the information: the message must move from specialized to non-specialized jargon in order to make information accessible to a broader audience. In this process of language transposition, it is essential to observe several aspects that directly influence the assimilation of content by the public (Fischhoff & Scheufele, 2013).

There is a long history of visual communications in science. Nonetheless, most of the time, this contribution has not focused on improving the connection with the unskilled audience. There are two main problems in this process: first, not considering the audience properly, and second, visual material is usually treated as a complement rather than as an integrated part of the project (Estrada & Davis, 2015). Plain Language deals with the whole process of language transposition, including, in addition to discursive aspects and textual formulation, these two issues.

The Plain Language Movement began to be structured in the 1970s, primarily in English-speaking countries, with the initial goal of improving access to justice and enabling consumers to make more informed decisions (Adler, 2012; Fischer, 2018). Following organized popular pressure and the creation of a variety of materials that indicated the benefits and discussed ways to make written and visual language more accessible, the Plain Language Movement began to gain support from the government, non-profit organizations, and groups of professionals related to communication. Thus, it established a set of practices that simplify reading and understanding of texts and documents and today composes communication guidelines in governments of different countries (Adler, 2012; de Oliveira & Bocchini, 2013; Fischer, 2018).

Design is an aspect to be considered by proposals according to the Plain Language Association International for effective and affordable communication. More specifically, Information Design seeks to adapt the representation of knowledge to cognitive processes in order to promote clarity. According to Subiela-Hernândez (2017), the idea of information design is linked to information architecture, information visualization, and eventually also related to infographics. A widely used definition of Information Design is that of Horn (2000, p. 15), which describes it as: "the art and science of preparing information so that it can be used efficiently and effectively by humans." Baer & Vacarra, (2008), in a similar direction, state that Information Design aims at "(...) translating complex, unorganized, or unstructured data into valuable, meaningful information."

Information Design has crossed disciplinary boundaries and blended disciplinary perspectives (Albers & Mazur, 2014; Cairo, 2012); it uses the principles of visual communication but it does not restrict itself to them to the extent that a designed text can contribute to effective communication. Thus, besides the use of textual and numeric elements, it also works with data plots, diagrams, illustrations, and maps; Information Design structures and organizes information elements that allow the user to find a way through the "maze of information" (Bonsiepe 1999, p. 59).

In this sense, Information Design and the Plain Language Movement (Mazur 2000; Petelin 2010) have high possibilities for connection, since they start from very similar assumptions. Both consider the concern with the human component (user) and the ergonomic aspects of communication as paramount and are guided by the idea that gave rise to the very concept of Clear Language: the recognition of the right that everyone has to understand relevant content for their daily lives, as is the case with scientific communication today. According to Scheufele and Fischhoff (2013), "we all need science to make effective decisions in our lives."

Scientists have endeavored to present clear and standardized reference systems for their respective fields of research. According to Tufte (1997, p. 10), those who make a discovery often need to present it, and representing, describing, and illustrating is also part of the process of building knowledge. Two good examples are the periodic table of elements (Figure 1) and the emblematic diagram of the Napoleonic March for Moscow, in which Charles Joseph Minard presented various numerical and statistical data on the disastrous losses of Napoleon's army during the campaign for Russia in 1812 (Figure 2).

Figure 1. Scerri, E. R. (1998). Mendeleev's periodic table of elements. (p. 82)

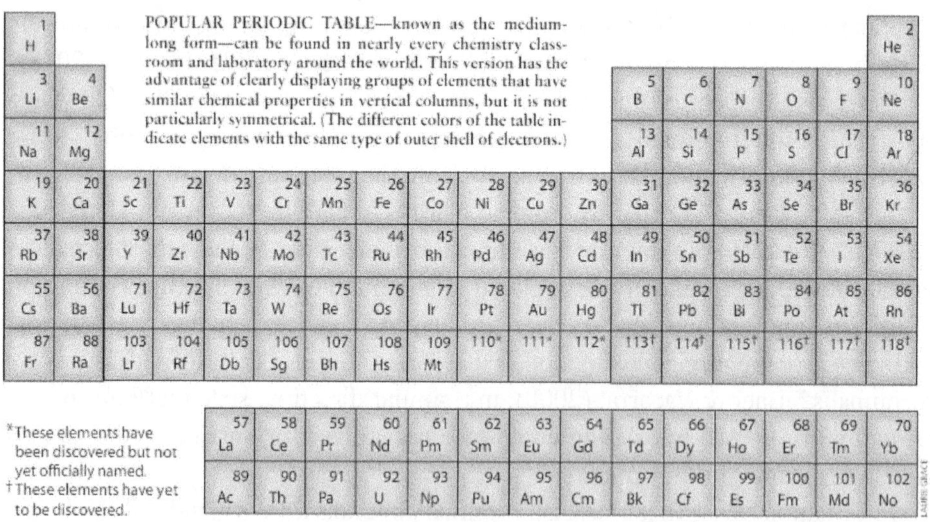

Figure 2. Minard (1861). Figurative Map of the successive losses in men of the French Army in the Russian campaign 1812–1813. (p.1)

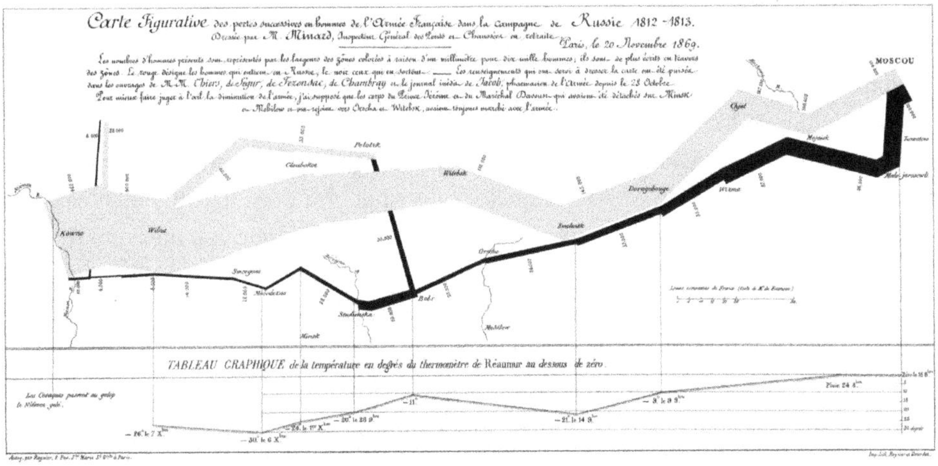

What these examples have in common is a lean and clear text coupled with the use of sign structuring strategies to make it easier to read a large amount of information. In the case of the periodic table, the elements are ordered simultaneously in a system of rows and columns, one of colors and one of spatial groupings, according to their chemical and physical properties. Minard's map also makes use of spatial and color relationships, adding a linear-temporal dimension and graphical features such as line thickness and line shape.

Thus, visual and diagrammatic systems have shown to be an excellent answer to the problem of data visualization (Franco & Borges, 2019; Franco 2017). One of the main reasons for this is very simple: Once the information reaches the retina, the brain first processes data such as shape, color, and orientation. Humans can interpret images in just 13 milliseconds, while texts are processed much more slowly (Potter, 2013, p. 275).

Studies have already shown that there is an imminent preference for receiving visual-static and visual-dynamic scientific information (static or animated images and diagrams) over texts and audios (Kaercher, 2015). Moreover, it is not just a matter of taste: when subjected to testing the content presented, the higher scores tend to occur among those who have received information through visual schemes.

Thus, information designed, organized, and clarified in visual-diagrammatic systems facilitates reading, assimilation, and use of information, as it visually reveals patterns and relationships that would not be easily deduced without graphic representations, which constitute "one of the basic sign systems conceived by the human mind" (Bertin, 1983).

PRINCIPLES OF INFORMATION DESIGN: A SEMIOTIC APPROACH

Information Design is not only interested in the design itself, but also in the information it conveys, which Baer & Vacarra (2008) calls content-focused. It focuses on the content of the message, which is what the reader or user needs to capture. Plain Language Association International claims in its recommendations that a planned communication "starts by considering what you want to achieve in writing communication. If it succeeds, what will happen? How might you measure that?"

To answer the above questions, it is necessary to think about who the audience is, what information they may already know, what they may need and what they want to achieve by reading the communication (PLAIN). In Information Design methodology, this is the user-centric mindset principle (Baer & Vacarra, 2008).

Information Design uses Graphic Design to make the message understandable to its target audience, but the end is not the design, it is the content, the message itself. If the design is attractive, but the reader cannot capture what is useful, this may be Graphic Design, but it is not Information Design. The strategies used to "select the best mix of the following language and design elements to employ" (PLAIN) constitute what Baer & Vacarra (2008) call tools of the trade.

Lohr (2003, p. 81) points out that, in terms of design in general, "decisions do not take a cookbook approach. Many factors influence the design.'' Pettersson (2010) further states that it is not possible to develop Information Design rules on how to best design a message and develop information sets. That is, for each specific case, the designer must be able to observe, analyze, and understand the problems to be solved and find the design solutions.

However, it is possible to formulate several effective and efficient general principles and guidelines (Pettersson 2010). Different authors divide and name design steps in different ways, but always guided by a similar scope. Several authors have offered guiding principles for design (Frascara, 2004; Lipton, 2011; Pettersson, 2010; Tufte, 1983). In general, there is something that guides these principles. Two primary Design's principles are the focus on the message recipient's questions and the message's formal issues.

From the framework of Charles Sanders Peirce's Semiotics (1839-1914), we propose to think of the message as a sign. According to Peirce (2012), a sign is anything that can represent something (an object) to someone and from that representation, produce a mental sign resulting from the interpretation, and that constitutes the understanding of the first sign. Peirce calls *semiosis* this sign produced by the understanding of interpretant and the process of interpreting signs (Santaella *et al.* 2016). In this case, the message of scientific dissemination would be the Sign that would represent the research and its results, so that a non-specialist audience produces interpretants that, to some degree, must correspond to the object (Figure 3).

Figure 3. Sign Concept applied to Popular Scientific Communication

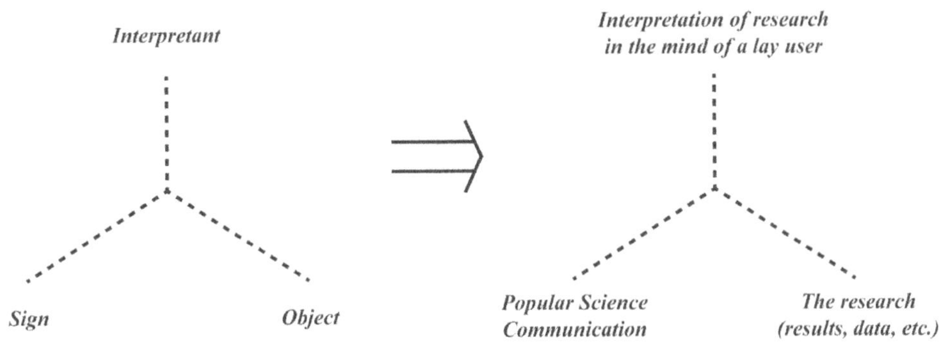

Peirce's triadic concept of sign enables the categorization of principles to facilitate their use by scientists and other professionals interested in improving their communication skills through Information Design. Thus, the guiding principles to be observed in the design of scientific communication messages can be categorized in three groups as follows:

The first group principles are focused on the formal constitution of the sign: they are related to the structure, composition, and shape of the sign itself, that is, the message. The colors, shapes, lines, thicknesses, textures, and other graphic features; legibility and readability, typography.

The second group principles refer to the correspondence between the sign (message) and what it represents (its object, which, in this case, is the research and its results). In what ways the message-sign can represent the object. What are the meanings of the formal aspects presented above, and which is the best way to represent them. The critical question is to assess whether the message (sign) matches your object (the search and its results). In semiotic terms, it relates to the sign and object relationship. The message must have a logical organization that embraces all its elements, creating a narrative that gives meaning to the information. It is essential to translate scientific jargon into easily understandable language.

The third group principles are focused on pragmatic-communicational principles, i.e., effects, practical consequences that could conceivably result from messages in a given context and audience. In semiotic terms, it relates sign, object, and produces a third sign, the interpretant, which would be the effect of the message. These principles note that signs can be produced from a practical interpretation of messages and stick to the practicalities of how to sign constructions. Their meanings should reach specific audiences in particular contexts. Pragmatic-communicational principles pay attention to effective ways to communicate, for this it is crucial to know the audience. It is essential that the planning has user focus. It is important to note that the principles above are interdependent and do not have a correct order of application.

GETTING BETTER AT INFORMATION DESIGN: A BRIEF CASE ANALYSIS

This topic will present cases of the use of Information Design principles in scientific communication, specifically in the field of Entomology. The choice for Entomology was made for two reasons: the first is that there is already a consolidated concern and research in the area regarding scientific communication (Berenbaum, 2017; Tribull, 2017). For example, the 2017 annual meeting of the Entomological Society of America (ESA) focused on scientific communication with insect scientists. It has been a priority for the Entomological Society of America since 2016 (Berenbaum, 2017).

The second reason is the fact that the field faces some resistance from the lay public. Kellert (1993) identifies in the lay public an aversion, fear, and negative attitudes related to invertebrates. Rayda Krell, Ph.D., a researcher at Western Connecticut State University, says "Do all entomologists know when you're at a party and someone says 'what do you do?' and you say 'I study bugs', people respond 'Eww, gross. I hate bugs!'" (Entomology Today, 2019).

The two cases chosen were infographics, schematic visual representations that integrate textual and image information to form a narrative, facilitating the reading of content and simplifying mental tasks. Infography is used continuously in the context of scientific communication, both in digital and print publications, as it is a type of graphic planning capable of arranging complex scientific data in a diagrammatic logic that makes the information clearer. According to Rajamanickam (2005), building the visual representation of information is not just merely like translating. "It involves filtering information, establishing relationships, differentiating patterns, and representing them in a way that allows the reader to understand that such information builds something meaningful" (Rajamanickam, 2005).

Pettersson (2010) states that verbal messages work well when the message content is analytical, detailed, logical, narrative, theoretical and sequential. Visual messages are preferred when the content is emotional, holistic, immediate, spatial, and visual. For complex messages, combining verbal and visual representations may be the best choice. Precisely because they are hybrid systems, infographics[1] are used continuously in the context of scientific dissemination, both in digital and print publications.

Case 1: Anatomy of the Bite by Rebecca Konte

This work (Figure 4) won the National Science Foundation (U.S.) Vizzies Challenge in 2018. It seeks to present how mosquitoes are capable of transmitting diseases that make them dangerous to human health. According to the author (Konte, 2019), "this piece intends to provide a lay audience with a visually targeted exploration of the biomechanics of mosquito feeding. By presenting this information clearly and engagingly, this piece allows an audience to understand how mosquitoes function as disease vectors."

Figure 4. Konte, Rebeca (2019). Anatomy of the Bite.

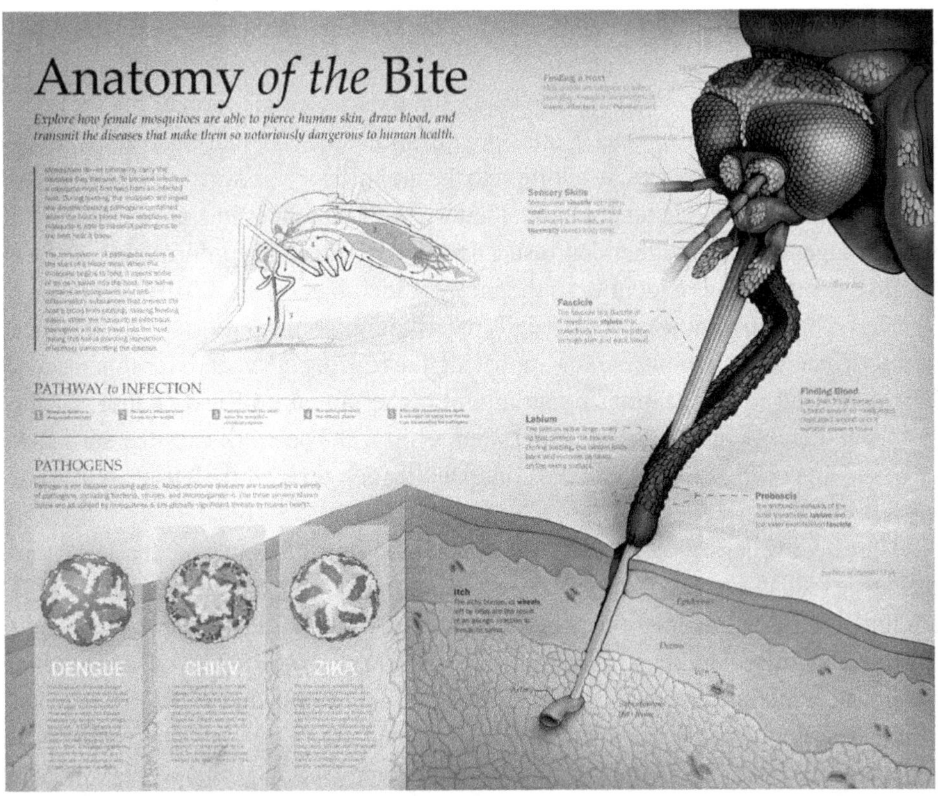

The first group principles are focused on the formal constitution of the sign. The work is an illustrated diagram, which brings a typical infographics composition. The title is in the upper left corner, followed by a brief text for contextualization, and the main image (which should immediately indicate to the reader the theme covered in the piece) is linked to topics arranged in a logical sequence. Other curiosities, or secondary information, appear at the bottom of the infographic.

This composition, or grid, the division that guides the elements of the page and the reader (Fetter, 2011), is traditional because of its ability to accommodate in a more orderly way a large volume of data, grouping, and (or) dividing them, creating a narrative logic. However, although the work follows this basic structure with few changes (for example, the main image is usually centered, and in this case, it is shifted to the right), the arrangement of the elements may not establish an intuitive reading order, and may be confusing to the reader.

The dimensions and formats of text blocks, images, and other graphic elements, such as lines and arrows, as well as their insertion in the composition, must always be planned to take into account the western reading direction: from left to right, from top to bottom. The elements of the composition must be arranged for this purpose in order to guide the reader's eyes. It configures a path that deviates from the intuitive eye route.

Information Design seeks to emphasize the most essential parts of the message through visual separation and spatial organization, such as margins, wires, bars, vignettes, frames or pictures, in addition to the graphic accentuation of the elements, such as the use of bold letters or capital letters, the differences in size, the highlight with stronger colors or higher contrast. The emphases make the information surface heterogeneous and give identity to the information object (Dur, 2014).

The typographic choices only partially meet the ergonomics criterion: the presence of a serif, which is responsible for improving the reading flow, and, for this reason, is widely used in publications containing a higher amount of text (Fetter, 2011). In this case, however, the usage criteria are reversed. The work uses the serif font only in some of the titles and in the contextualization text, while it uses in most of the texts a non-serif font.

The effectiveness of the message is further compromised by two other aspects that influence readability and legibility: size and contrast. Being a piece of considerable amount of written information, the presentation format requires a larger dimension for the texts to be legible with a minimum of visual comfort. The background of the image, a mixture of colors and textures little identifiable due to the low opacity, pollute the composition and disturb the contrast with the text, making it difficult to recognize the characters. The "dirtiness" of the composition might be explained as an excess of signs, unnecessary information that ends up more confusing than adding (Dur, 2014). Such excess makes it difficult to establish a primary focus of attention, impairing concentration and the capture of information (Karakas, Manisaligil, & Sarigollu, 2015).

The illustrations, a determining factor in the "appeal that a publication exerts on its reader" (Fetter, 2011, p. 66) arouse the reader's interest due to the considerable similarity with its real correspondent; presenting with relative clarity of details the colors and textures of the insect. Despite this, the images have indistinct edges, as they present a low gradation of opacity, which makes them blend into the background — some of the accompanying texts have the same problem.

As for the fidelity of the representation of the infographic concerning the data and analyses of the scientific universe it represents (correspondence between the sign and its object), the piece reasonably achieves its objective - "above all, show the data" (Tufte, 2007, p. 92). Despite the formal problems described above, in the appropriate format, the piece would fulfill, with some efficiency, the role of facilitating data communication.

Regarding the pragmatic-communicational principles, the project details described by the author (Konte, 2019) state that their audience is the lay audience, a somewhat broad target audience. As Khoury et al. (2019) show, when scientists publicize their work, they usually want to communicate with "the general public." However, different audiences require different approaches and, consequently, different discourses, languages, and graphic projects. In this sense, the definition of the target audience must be as specific as possible.

As already discussed, for both plain language and information design, knowing and listening to the target audience is essential. "It is also necessary to consider the needs and aspirations, placing these recipients at the beginning of the production process, based on their daily reality" (de Oliveira & Bocchini, 2013). The project details also deal with the intended final presentation format: print. The text does not contain information on the format, details on paper, and printing or distribution of the material. The internet does not offer the necessary image quality or dimension for a proper reading of the information.

Initiatives like this are essential to promote the dissemination of science. However, the examination conducted here reinforces the need for cooperation between designers and scientific disseminators to design more efficient communication plans.

Case 2: Species Hunt by Stegmaier, Alexander; NGM Staff; 5W Infographics — National Geographic Magazine

This example (figure 5) is an infographic that shows the number of animal species discovered and recorded as of the date of publication, compared to the number of species that were estimated to exist on the planet. According to the authors:

[...] scientists have cataloged more than 1.7 million species. But there's still a lot of work to be done. Estimates of the total number of species range from 3 million

to 100 million; one new study puts the figure at 8.7 million, give or take a million. The numbers at right, compiled from many surveys and statistical projections, offer the best guess for species in selected categories. As these examples illustrate, most vertebrates have been identified, but many invertebrates — especially insects — remain for future generations of explorers to discover (Stegmaier et. al. 2013).

Figure 5. Stegmaier et. al. (2013). Species Hunt.

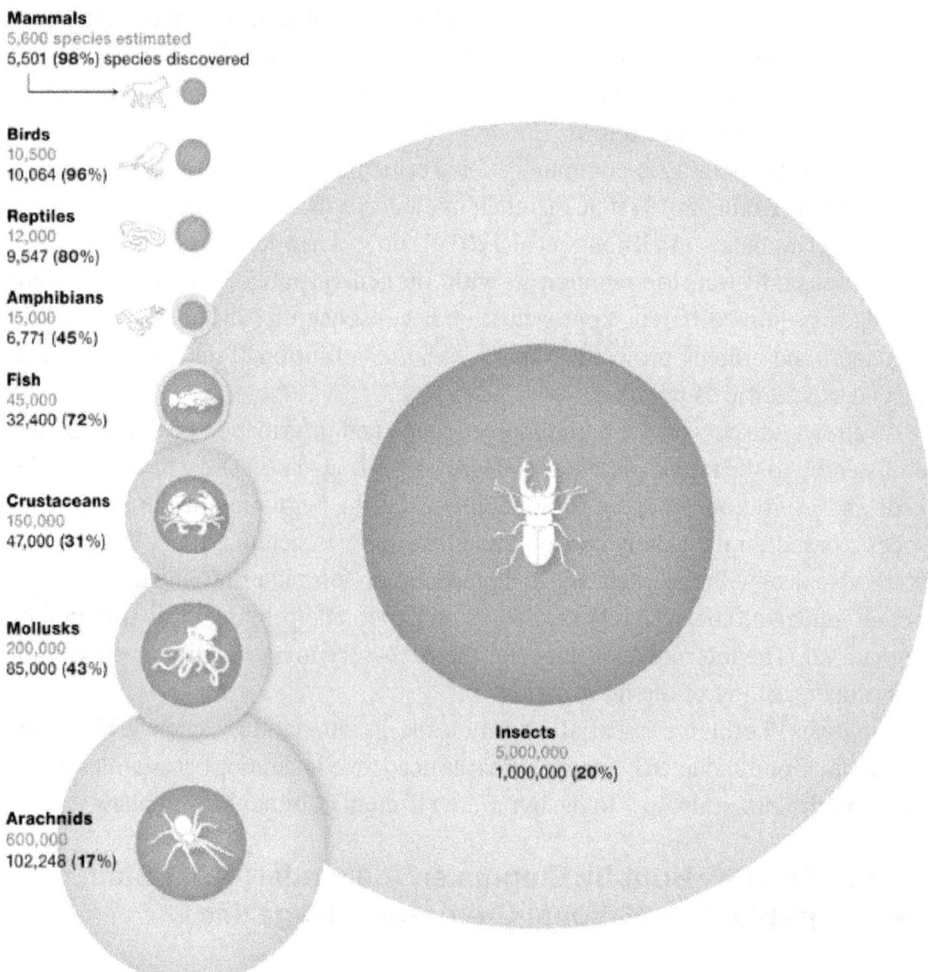

This perception is helped by the presence of the circles in which the illustrations are inserted, whose size and color are directly related to the numbers that the research brings. The darker inner circles visually represent the number of species discovered and recorded, belonging to the phylum or class in question. Outer circles, in turn, refer to the estimated number of existing species. The colors of the circles follow the colors of their respective textual information. This type of representation, part of a whole, illustrates percentage data well, making information that would otherwise be abstract quite clear.

The illustrations of the representatives of the phyla and classes are pregnant[2] because, in addition to being constant images in the popular imagination, they contain few details, are practically flat, and have well-defined silhouettes, in order to be easily identifiable, even in small sizes.

The work has only one font, simple and without a serif, which is appropriate since there are no large blocks of text: the information is extremely lean. The only variation, besides the color, is the weight; the titles and percentages appear with greater weight, in bold, due to the emphasis given to this information. Although the work uses colors as information, the hue is not the main variant, but the contrast between them, so that the piece works even in shades of gray.

The analysis demonstrates proper use of analogical strategies and clear language: the infographic presents clearly and succinctly, in both verbal and non-verbal signs, the information it proposes to communicate. It is interesting to note that this work refers to a much smaller collection of data compared to the first example, which directly influences the synthetic quality of the final result; filtering and compiling data is a fundamental step in the process. The work is assertive in the fidelity of the sign's representation to the data and analysis of the scientific universe it represents.

On the pragmatic-communicational aspects, it is essential to note that the National Geographic Magazine is a respected and old scientific publication, which, today, reaches mainly a young audience and students. According to its media information kit, "a third of National Geographic Magazine readers are millennials [...] which makes millennials the largest audience segment of our readership". This new generation, "may face three major problems about management learning: lack of concentration, lack of engagement, and lack of socialization" (Karakas, Manisaligil, & Sarigollu, 2015). It is not by chance that, over the years, there has been a general movement towards the "cleaning" of graphic compositions, which accompanies the demand for leaner and more objective texts (Fischer, 2018). In this sense, information design meets the demand for a well-defined public profile.

CONCLUSION

In a context of intense circulation of large amounts of information, Information Design presents itself as an essential communication tool. Communication planned by Information Design, in addition to bringing content modeled to reduce information overload, visually reveals patterns and relationships that are not comprehensible easily without graphical representations. In this way, it enables clarity of communication and can help to disseminate the thinking, processes, and results of science to broader and non-specialist audiences, especially Lifelong Learners.

As the Chapter intended to demonstrate, Information Design, combined with plain language, has the potential to increase the attractiveness, the ability to understand and, consequently, the efficiency of communication. Increasing the potential for dialogue and the visibility of science is especially essential at the moment when there is a growing worldwide influence of anti-scientific voices active in public discourse (Hulcr et al., 2019).

Formal scientific communication is appropriate to provide a detailed record of research and discovery. However, it is not very attractive to the unskilled reader. One aspect that the research still seeks to develop is related to the relationship between storytelling techniques and Information Design. Storytelling visually can be a powerful tool for involving humans with science. A future direction for research concerns the understanding of the narrative as an essential component of the meaning in Information Design. Concerning science communication, the research also aims, in the future, to systematize the principles of Information Design in a visual booklet, to facilitate their use by scientists and other professionals interested in improving their communication skills through this area.

ACKNOWLEDGMENT

We kindly thank Rafael Frasson Cecato Cabral for the English revision of the manuscript and Dr. Juliana de Paula-Souza (Biological Sciences Center, Federal University of Santa Catarina) for the general review and proofread.

REFERENCES

Adler, M. (2012). The plain language movement. In The Oxford handbook of language and law. Oxford University Press. doi:10.1093/oxfordhb/9780199572120.013.0006

Albers, M. J., & Mazur, M. B. (Eds.). (2014). *Content and complexity: information design in technical communication*. Routledge. doi:10.4324/9781410607409

Baer, K., & Vacarra, J. (2008). *Information design workbook: Graphic approaches, solutions, and inspiration+ 30 case studies*. Rockport Publishers.

Berenbaum, M. R. (2017). Communicating about science communication: A brief entomological history. *Annals of the Entomological Society of America, 110*(5), 435–438. doi:10.1093/aesaax060

Bertin, J. (1983). *Semiology of graphics* (W. J. Berg, Trans.). The University of Wisconsin Press.

Bonsiepe, G. (1999). Interface: An Approach to Design: In Memoriam Jovita (1941-1998). Jan van Eyck Akademie, Department of Design.

Brockington, G., & Mesquita, L. (2016). The consequences of bad science communication. *Revista da Biologia, 15*(1), 29–34. doi:10.7594/revbio.15.01.03

Bucchi, M., & Saracino, B. (2016). "Visual Science Literacy" Images and Public Understanding of Science in the Digital Age. *Science Communication, 38*(6), 812–819. doi:10.1177/1075547016677833

Cairo, A. (2012). *The Functional Art: An introduction to information graphics and visualization*. New Riders.

de Oliveira, L. L., & Bocchini, M. O. (2013). *Associação entre design da informação e linguagem simples em materiais informativos das prefeituras de São Paulo (1990), Campinas (2000) e Várzea Paulista (2009)*. Retrieved November 16, 2016, from http://pdf.blucher.com.br.s3-sa-east-1.amazonaws.com/designproceedings/cidi/CIDI-4.pdf

Dur, B. I. U. (2014). Data visualization and infographics in visual communication design education at the age of information. *Journal of Arts and Humanities, 3*(5), 39–50.

Estrada, F. C. R., & Davis, L. S. (2015). Improving visual communication of science through the incorporation of graphic design theories and practices into science communication. *Science Communication, 37*(1), 140–148. doi:10.1177/1075547014562914

Fetter, L. C. (2011). *Revistas, design editorial e retórica tipográfica: a experiência da revista Trip* (Unpublished doctoral dissertation). Retrieved November 20, 2019, from https://www.lume.ufrgs.br/handle/10183/30193

Fischer, H. (2018). *Clareza em textos de e-gov: uma questão de cidadania*. Com Clareza.

Fischhoff, B., & Scheufele, D. A. (2013). The science of science communication. *Proceedings of the National Academy of Sciences of the United States of America, 110*(Supplement 3), 14031–14032. doi:10.1073/pnas.1312080110 PMID:23942127

Franco, J. R. (2017). O Design de Informação e a narrativa diagramática em Whale Hunt de Jonathan Harris. *METAgraphias, 2*(3).

Franco, J. R., & Borges, P. M. (2019). Diagrammatic relations of probative strength and inferential progression through semiotics. *Semiotica, 2019*(228), 77–89. doi:10.1515em-2018-0088

Frascara, J. (2004). *Communication design: principles, methods, and practice.* Allworth Communications, Inc.

Gross, A. G., Harmon, J. E., Reidy, M., & Reidy, M. S. (2002). *Communicating science: The scientific article from the 17th century to the present.* Oxford University Press on Demand.

Horn, R. E. (2000). Information design: Emergence of a new profession. In R. Jacobson (Ed.), *Information design.* MIT Press.

Hulcr, J., Dharampal, P. S., Hamm, R. L., Pearson, G., & Gibson, C. (2019). Influence Is Power: Strategic Communication for Entomologists. *American Entomologist (Lanham, Md.), 65*(2), 92–96. doi:10.1093/ae/tmz024

Kaercher, A. L. (2015). *Adaptação automática de interface de acordo com características do usuário.* UniRitter. Retrieved December 2, 2019, from https://bit.ly/35zrL8K

Karakas, F., Manisaligil, A., & Sarigollu, E. (2015). Management learning at the speed of life: Designing reflective, creative, and collaborative spaces for millenials. *International Journal of Management Education, 13*(3), 237–248. doi:10.1016/j.ijme.2015.07.001

Kellert, S. R. (1993). Values and perceptions of invertebrates. *Conservation Biology, 7*(4), 845–855. doi:10.1046/j.1523-1739.1993.740845.x

Khoury, C. K., Kisel, Y., Kantar, M., Barber, E., Ricciardi, V., Klirs, C., & Valiño, Á. (2019). Science–graphic art partnerships to increase research impact. *Communications Biology, 2*(1), 1–5. doi:10.103842003-019-0516-1 PMID:31396575

Konte, R. (2019). *Anatomy of a Bite. Project Description.* Retrieved December 3, 2019, from https://meetings.ami.org/2017/project/5-2/

Krause, K. (2017). A framework for visual communication at Nature. *Public Understanding of Science (Bristol, England), 26*(1), 15–24. doi:10.1177/0963662516640966 PMID:27117485

Lipton, R. (2011). *The practical guide to information design.* John Wiley & Sons.

Lohr, L. L. (2007). *Creating graphics for learning and performance: Lessons in visual literacy*. Prentice Hall Press.

Mazur, B. (2000). Revisiting plain language. *Technical Communication (Washington)*, *47*(2), 205–211.

Minard, C. J. (1861). *Tableaux graphiques et cartes figuratives, Bibliothèque numérique patrimoniale des ponts et chaussées*. Retrieved December 13, 2019, from https://patrimoine.enpc.fr/document/ENPC01_Fol_10975

Peirce, C. S. (2012). *Charles S. Peirce, Selected Writings*. Courier Corporation.

Petelin, R. (2010). Considering plain language: Issues and initiatives. *Corporate Communications*, *15*(2), 205–216. doi:10.1108/13563281011037964

Pettersson, R. (2010). Information design–principles and guidelines. *Journal of Visual Literacy*, *29*(2), 167–182. doi:10.1080/23796529.2010.11674679

Plain Language Association International (PLAIN). (n.d.). *What is Plain Language?* Retrieved December 10, 2019, from https://plainlanguagenetwork.org/plain-language/what-is-plain-language/

Potter, M. C., Wyble, B., Hagmann, C. E., & McCourt, E. S. (2014). Detecting meaning in RSVP at 13 ms per picture. *Attention, Perception & Psychophysics*, *76*(2), 270–279. doi:10.375813414-013-0605-z PMID:24374558

Rajamanickam, V. (2005). Infographics: Handout. National Institute of Design, 1-14.

Santaella, L., Franco, J. D. O. R., De Bona, C. M., & Espínola, V. (2016). Por uma semiótica não-tergiversante: Análise do site Conductor-MTA. me. *LÍBERO*, (28), 37–76.

Scerri, E. R. (1998). The evolution of the periodic system. *Scientific American*, *279*(3), 78–83. doi:10.1038cientificamerican0998-78

Stegmaier, A. (2013). Species Hunt. *National Geographic Magazine*. Retrieved December 13, 2019, from https://www.nationalgeographic.com/magazine/2013/04/species-hunt/

Subiela-Hernández, B.-J. (2017). Diseño de la información y progreso social: Prospectiva, perspectiva y retos. *El Profesional de la Información*, *26*(6), 1019–1024. doi:10.3145/epi.2017.nov.01

Today, E. (2019). *Why Science Communication Ought to Be Part of the Job for Entomologists*. Retrieved December 13, 2019, from https://entomologytoday.org/2017/02/01/why-science-communication-ought-to-be-part-of-the-job-for-entomologists/

Tribull, C. M. (2017). Sequential science: A guide to communication through comics. *Annals of the Entomological Society of America*, *110*(5), 457–466. doi:10.1093/aesaax046

Tufte, E. R. (2007). *The visual display of quantitative information*. Graphics Press.

Tufte, E. R., & Robins, D. (1997). *Visual explanations*. Graphics Press.

ENDNOTES

1 Infographics and data visualization, as visual-diagrammatic schemes, are two possibilities for the application of Information Design (Dur, 2014).

2 The concept of pregnancy is based on the idea of structural coherence of a shape and therefore on the sense of symmetry, order, regularity, and simplicity.

Section 2
Communities of Learning

Chapter 5
Presentation of a Social Responsibility Project on Social Media and MOOCs Within the Scope of Scientific Communication:
Basic Health Knowledge, First Aid, and Coronavirus Example

Gökhan Kuş
Anadolu University, Turkey

ABSTRACT

The learning activities of individuals are not only limited to school, but they also continue learning in all areas of life. One of the most effective ways of achieving this in the evolving socio-cultural and technological environment is the social media environment. The authors addressed the issues separately for the deficiencies identified in the study. They offered the trainings they have done in various platforms.

INTRODUCTION

Human beings continue their lives through learning from the moment they are born. The learning activities of individuals are not only limited to school, but they also continue learning in all areas of life. Gaining knowledge and experience in these

DOI: 10.4018/978-1-7998-4534-8.ch005

structured or unstructured learning activities, especially in terms of health, enables the development of conscious behavior in case of an incident which may require first aid intervention.

Rapid development in technology has also transformed learning. It is important for people who spend time in online environments, especially on social media, to have access to useful information during these periods. Providing individuals with information on social media in an interesting and simple way is important in terms of making the masses conscious.

LIFELONG LEARNING

Lifelong education, or lifelong learning, which could be expressed as an educational phenomenon resulting from continuous needs of individuals, refers to an understanding which includes an individual's acquiring knowledge, skills, attitudes and habits by making use of all available learning environments in line with his/her goals and needs so as to improve himself/herself (Lengrand, 1989; Cassin, 2002; Budak, 2014; FREREF, 2004; Commission Européenne, 2008). In the commission of UNESCO on the development of education that convened in 1972, it was decided that education is a lifelong activity (Lee and Friedrich, 2011). In this study, the concept of lifelong learning will be used.

Differently from other educational structures, in lifelong learning, an approach which places the individual at the very center is adopted. Other differences of lifelong learning from other structures can be listed as an emphasis on informal learning, the change in the role of the school, decreasing the weight of the state in education, strengthening the role of social stakeholders and a special emphasis on the necessity of not limiting education to a certain period of time. Wain (1993) names this philosophy as novel humanism and argues that humanism is the most important element in the formation of the concept of lifelong learning (Murphy, 2000).

The EU commission conducted a study titled (Towards a Learning Society" in 1995, and upon the decision of the commission, 1996 was accepted as the year of lifelong learning in Europe. It is possible to see goals and practices related to lifelong learning in other educational projects of the EU. For instance, some of the pillars of the Leonardo da Vinci program run in the field of occupational education are related to lifelong learning (Akbaş and Özdemir, 2002); Kozikoğlu and Altunova, 2018). Moreover, one of the priorities of the program is to contribute to lifelong learning by using tools associated with information society (Duman, 2001). In accordance with 2020 objectives, the EU has established a de facto educational priority by integrating the policies and measures it has designed throughout years (Göksan, Uzundurukan and Keskin, 2009; Sağlam, Özüdoğru and Çıray, 2011; Toprak and Erdoğan, 2012).

There are six basic dimensions of lifelong learning determined by The EU commission, which are multiple participation, learning by using information and communication technologies, qualification of the resources, ease of access to learning environments, creation of a learning culture, establishing a set of criteria for determining the level of learning (Aktan, 2007; Commission Eurpéenne, 2008; Özmususl, 2018).

SOCIAL RESPONSIBILITY PROJECTS

The term "Corporate Social Responsibility" (CSR) has been commonly used since the 1970s. This concept has spread all over the world especially through various activities carried out by internationally or globally recognized corporations.

Social responsibility, which is about the behavior of the business towards its social circle and individuals, has a philosophy of giving back what it has received from the society and upholding the benefits of the society in its activities. The corporate social responsibility understanding has an important place for business in promoting the products and services they produce to the market in the best possible way and strengthening their brand image (Aktan and Börü, 2007). There is an increasing tendency to buy the products and services provided by companies which make this understanding an essential part of their activities and thus improve their brand image (Özdemir Çakır, 2009).

State institutions, which are responsible for the education of the society must determine the lacking points in society in terms of current technologies and educational environments and do the necessary works in order to fill these gaps. N this study, since we believe to be deficiencies in these areas, we will handle the subjects of "Basic Health Knowledge and First Aid" and "corona virus" which has become a current issue as of the end of 2019 and the beginning of 2020.

In the evaluations made as a result of various studies conducted on a national and international scale, it has been determined that individuals from different walks of life have very low levels of information about basic health and first aid knowledge (Kuş, 2018). There is a fact that has come out as a result of these studies. Individuals themselves are not aware of the fact that they have deficiencies regarding basic health and first aid knowledge. These deficiencies cannot be eliminated due to individuals' not being able to spare time as a result of their working hours and physical factors. At this point, it is the corporate social responsibility projects that should come into play.

A new health problem that has affected the whole world has appeared as of late 2019 and early 2020: a corona virus known as COVID19. Corona virus mainly affects respiratory system. However, it does not cause the same damage in all people.

When we look at the epidemics in which this virus was active in the historical process, we can see the recent SARS and MERS epidemics. Therefore, we will be faced with coronavirus and similar viruses as long as humanity exists. Accordingly, in the context of coronavirus, we need to raise the awareness of individuals about diseases and epidemics originating from viruses and bacteria and keep them away from a panic environment that might unnecessarily develop. Another important element in taking measures against diseases is to get information about developing processes from reliable sources. Unfortunately, in cases of diseases and epidemics, some people share unfounded information, whether on purpose or to have fun, that would cause public unrest and create panic and anxiety in people in the Internet environment. As in other basic health related issues, people should be provided with correct and timely information. One of the most effective ways of achieving this in the evolving socio-cultural and technological environment is the social media environment. Everyone should be given correct and effective information in a timely manner through sharing with rich content to be prepared by official institutions or relevant authorities.

Given that we live in an age of technology, one of the important issues that have come to the forefront is new communication technologies. Especially in recent times, communication models developing as a result of the Internet and internet-based technologies have had enormous effects on our daily and even hourly lives. Face-to-face interactions between people have decreased, and a variety of interactions and communication forms accessed through the Internet have appeared. Although these interactions were considered from time to time as complementing, supporting and developing one another, there are also negative evaluations stating that they are obstructing, conflicting and restricting each other. Thus, we can evaluate an incident that affects the whole society in terms of public health in daily life as a positive and negative example as follows (Arklan, 2016):

1. Information sharing over the Internet by official public institutions and organizations or relevant and authorized units diminishes the panic in people and guide people to the correct path.
2. Sharing of false, unnecessary tampered voice and video recordings over the Internet by individuals either consciously or unconsciously in order to have fun deeply affect daily lives and occasionally lead to public unrest depending on the significance of the subject.

Among the fields where relations in social life are mainly carried into virtual environment is the social media. On social media,

1. Sections or interactions from daily lives of individuals

2. Activities they do with their friends
3. Information about educational life
4. Events concerning everyone individually or socially

are shared necessarily or unnecessarily and correctly or incorrectly. What is important here is to use the social media effectively in accordance with one's purpose and area of interest. Through social media, public institutions and private corporations reach their clients and customers more easily and with low cost. Since this fact has been validated in many studies, many institutions have begun to invest largely in social media and similar service fields. Those institutions which have achieved this aim have the chance to inquire and learn about the needs and expectations of their target audience by knowing them more closely. The use of social media in education is also important in this sense. Especially in educational institutions such as university, the interaction between the user/student and the educational institution is increased, and both the institution and the user can share their opinion (request, information, criticizing etc.) in a very fast and effective way. As interaction intensifies, the users' commitment and loyalty to their educational institutions increase and thus a strong and serious relationship is established. On the other hand, if the institution does not share the necessary and correct information on social media effectively in a timely manner, the opposite situation occurs in which the users lose their sense of belonging and cut their relationship with the institution (Arklan, 2016; Aydın, 2017).

BASIC HEALTH KNOWLEDGE, FIRST AID AND CORONAVIRUS

Approximately how many cells are there in our bodies?
Where is the heart located in my body?
How many times does my heart beat in a minute?
What happens if my heart stops beating?
What happens each time my heart beats? Where does that sound come from?
How much blood do I have in my body?
What is in the blood?
Where does my heart send the blood?
What carries the blood to the heart?
How many different types of blood vessels do I have?
What is the function of the arteries?
What is the function of the veins?
How many times do I breathe in a minute?
Does my heart rate vary?

What is the function of oxygen?
How do the cells in my body receive oxygen?
What carries the oxygen in the blood?
What is an erythrocyte? What is its ideal number in the blood?
What is a leukocyte? What is its ideal number in the blood?
What is haemoglobin?
I had a blood test. There are lots of abbreviations on the test. What do they mean?
I feel very cold. Is my blood level low? Am I sick?
What is coronavirus affecting the world since December 2019?
How do you get infected by coronavirus and what are the symptoms?
What should or shouldn't I do in order not to catch coronavirus?

…

You constantly ask yourselves such questions in your daily lives or hear other people asking. You seek answers to these question from people around you. In order to have access to the correct answers to these questions or to be able to answer these basic questions, people should have information about health knowledge, first aid and coronavirus, at least at a basic level.

The World Health Organization defines health as "not only the absence of a disease or disability, but a total wellbeing in terms of physical, mental and social health". When we evaluate health form a physical perspective, keeping the internal environment of an organism stable in spite of changing internal and external conditions is called homeostasis. In other words, it is maintaining the body in a state of equilibrium. An organism can stay healthy to the extent that it can adapt to the stimulants it faces, stress, and internal and external changes. If it cannot adapt, a disease or other pathological situations may occur. In order to achieve such adaptation, it is necessary to have a healthy body, a healthy social environment and a healthy mental state. Health is a means to daily life, not the purpose of life (Billman, 2020). As for the factors affecting health, they can be categorized under three headings (Da Silva, 2018; Öztürk, 2019):

1. Individual factors: genetic structure, race, age, gender, immune system, and health history come to the forefront. These factors are those which we cannot intervene and change from outside.
2. Lifestyle or behavioural characteristics: This category includes physical activity, doing sports/exercise, adequate and balanced diet, and not using harmful substances such as tobacco and alcohol. These factors are related to an individual's lifestyle. They can also be labelled as changeable or correctable factors.

3. Environmental factors: This category involves the physical environment, chemical environment, biological environment and socio-cultural environment in which an individual lives, which directly or indirectly affects him/her. The living conditions of an individual, his/her occupation, educational level and family structure are among the leading elements in this category.

To put it simply, it can be said that individuals who can keep these three factors under their control always hold on to life better and maintain their lives in a healthy way. However, the essential requirement for this to happen is that individuals should be able to notice the importance of health and effective factors. For this purpose, they should use new communication technologies, especially social media, effectively.

One of the important subjects, the education of which can be provided through new communication technologies, basic first aid knowledge. When we think of the possibility of negative situation occurring in terms of our health or others' health anytime and anywhere, we can better understand that we all should know what we can and should do in such situations even at a basic level. Imagine a situation in which someone in front of you is fighting for his/her life and you have no idea what to do. What an unfortunate situation! First aid knowledge is an issue that everyone should know about at least at a basic level. Therefore, it is necessary to increase the knowledge and skill level of the society regarding first aid. However, providing face-to-face first aid education for everyone is costly both in terms of money and time. For this reason, new communication technologies, particularly social media, can be used to offer first aid education (Kuş, 2018).

One of the most frequent issues as regard first aid in our daily lives is related to airway obstructions. When we speak of airway, we think of mouth, nose, oesophagus, and trachea. Especially, in the presence of a foreign substance in the trachea, an individual experiences difficulty in breathing; if the trachea is completely blocked, the person cannot speak and starts to turn blue. As a result, if no intervention is made, the person may lose his/her life. Knowing what one can do in such situations which are frequently seen in daily lives and intervening in the correct way can help save lives. It also helps prevent wrong interventions. We should always keep this in mind: as a result of unnecessary or wrong interventions, people can suffer disabilities or even lose their lives. This situation which occasionally develops in adults is frequently observed in children and babies. Airway obstruction interventions in children and adults differ from the interventions in babies. Therefore, we need to know what to do when faced with such situations. It is important that short educational videos should be prepared on these issues and provided in social media environments (Rodriguez, 2020).

Another issue is related to blood and bleeding. A healthy adult individual (170 cm, 70 kg) has approximately 5 litres of blood circulating in his/her body. Blood is an essential fluid for life. Blood should not be considered just as a fluid. It includes all kinds of materials necessary for life. Oxygen, nutrients, organic and inorganic materials, water etc. are indispensably vital materials. All cells and tissues and organs formed by these cells must continuously be fed, and waste material must be removed from tissues and organs. These functions are fulfilled by the circulatory system composed of blood, vessels and heart. If a person loses 10% of his/her blood for any reason through bleeding, and if the bleeding cannot be stopped, the vital functions of the person are put under great risk. For this reason, bleeding should immediately be stopped. The timely and effective intervention in external bleedings in the body will increase the chances of that person to stay alive. So, it is highly important to inform the society about bleeding, which is one of the basic topics of first aid (Ural, 2017). This issue should also be covered by way of short video clips to be prepared by experts and be disseminated over social media.

There have been studies on coronavirus, a.k.a. COVID19, which was declared by the WHO as a pandemic, over the social media. Explanations made by public institutions and authorities attracted quite intense attention of the public on social media. The whole world has become alerted against this virus, which especially creates heavy clinical complications in the respiratory systems of risk groups and leads to death in people with chronic diseases. A lot of information has been shared on social media, which has been one of the world's common communication networks, whether they were correct/incorrect or real/false. Such information has reached millions of people through re-sharing of this information and comments made by other people. Health-related issues, about which people are quite sensitive, can cause concern and panic in people. Therefore, educational videos to be prepared by experts on the subjects and contents to be formed by official institutions and organizations or relevant authorities should be provided in social media environment, and thus people should be informed correctly. This is very important in terms of prevention and spread of diseases (Til, 2020; Zu, 2020).

SOCIAL MEDIA

In today's changing social structure, in addition to face-to-face communication between people, establishing communication directly or through virtual environments has gained importance. Social media use stands out among these new communication technologies. Social media terms simply refer to all types of interactions, dialogues and shares between people. Social media continues its development with different areas of use and different tools. The leading social media tools of today are Facebook, Twitter, WordPress, LinkedIn, Google, Foursquare, Blogger and YouTube (Usta, Uğur, Okur and Kuş, 2019). Social media has now been a part of learning processes.

The flexibility and ease of use of well-designed social media environments increase the number of social media users. In recent years, social media environments are increasingly seen as educational environments. By providing people with not only health issues but also subjects related to many areas in social media environments in a clear and understandable way, more people can be addressed. The increased use of social media environments by public institutions will increase access and sharing of information. The most important feature of social media is that it allows the followers to contribute to the content, to create their own contents and to make evaluations-suggestions- criticisms (Scott, 2011).

In 2016, it was decided to use social media in order to move interaction with students to an upper level, and for this purpose, Open Education System Facebook page was created. In November 2018, an Instagram account was created and re-structured. According to validated data, as of April 2020, the Facebook page has more than 284,000 followers, while 132,000 users follow the Instagram account.

EXPLAIN IN 90 SECONDS PROJECT

One of the social responsibility projects of Anadolu University Open Education Faculty is the project of "Explain in 90 Seconds" in the area of health. Within the scope of this projects, educational videos have been prepared initially for the followers of the Open Education System Facebook account and later expanded to the Instagram account followers on the following topics:

1. First aid in airway obstructions in adults
2. First aid in airway obstructions in babies
3. First aid in bleedings
4. What is coronavirus? How does it infect people) How should people protect themselves from coronavirus?

In line with the scenarios prepared on these topics, video shootings were carried out in green curtain studio environment. In these videos, the information was explained as simply as one who has no knowledge could understand, and for interventional practices humans or mannequins were used.

The duration of the videos shot was kept around 90-360 seconds, the videos were edited to be 90-100 seconds, and regarding current social issues, the duration of the videos were made longer. A significant number of views and sharing was achieved in a short time on this platform, which was followed by approximately 285,000. Screen shots related to this issue are presented below.

The video on airway obstruction in adults was shared on the Facebook page in three different times, and a total of 79,016 individuals was accessed.

Figure 1. The current viewing, likes, shares and comments on the video on airway obstruction in adults - Facebook

The same video was shared on the Instagram account once and 5,149 individuals were accessed.

Figure 2. The current viewing, likes, shares and comments on the video on airway obstruction in adults - Instagram

Information about the video on airway obstruction in babies is presented in the following figures.

Figure 3. The current viewing, likes, shares and comments on the video on airway obstruction in babies - Facebook

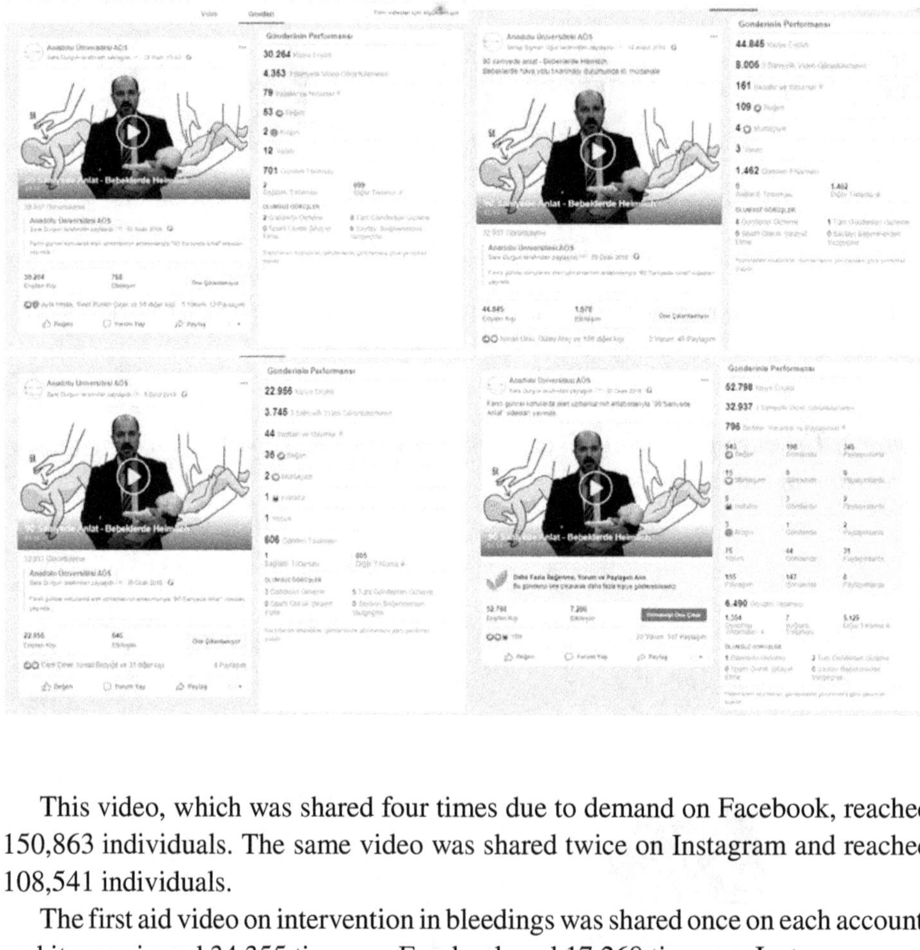

This video, which was shared four times due to demand on Facebook, reached 150,863 individuals. The same video was shared twice on Instagram and reached 108,541 individuals.

The first aid video on intervention in bleedings was shared once on each account, and it was viewed 34,355 times on Facebook and 17,268 times on Instagram.

Finally, two 90-second informative videos on COVID-19, which is a global pandemic, were prepared, but in order not to distract viewers' attention in this period of crisis, they were presented as one video. This video reached 39,667 individuals on Facebook and 55,609 people on Instagram.

Figure 4. The current viewing, likes, shares and comments on the video on airway obstruction in adults - Instagram

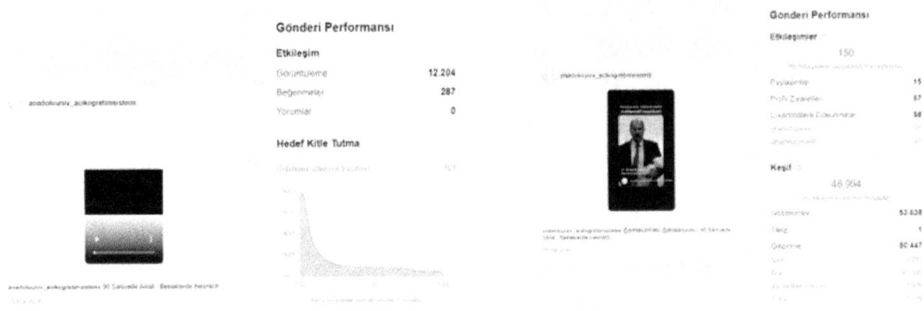

Figure 5. The current viewing, likes, shares and comments on the video on intervention in bleedings – Facebook and Instagram

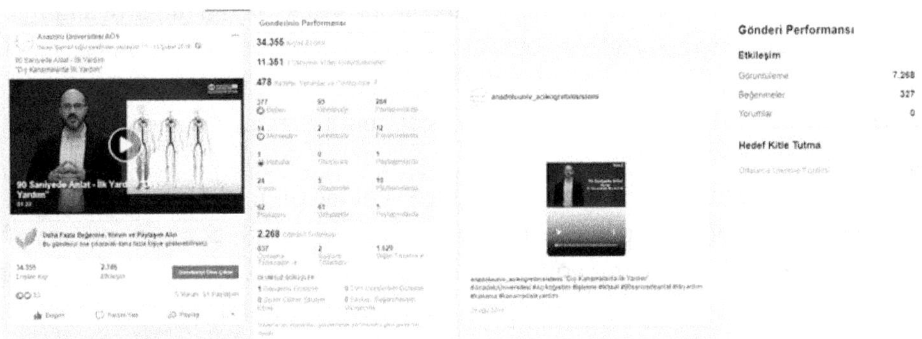

Figure 6. The current viewing, likes, shares and comments on the video on What is coronavirus? How does it infect people? How should people protect themselves from coronavirus? – Facebook and Instagram

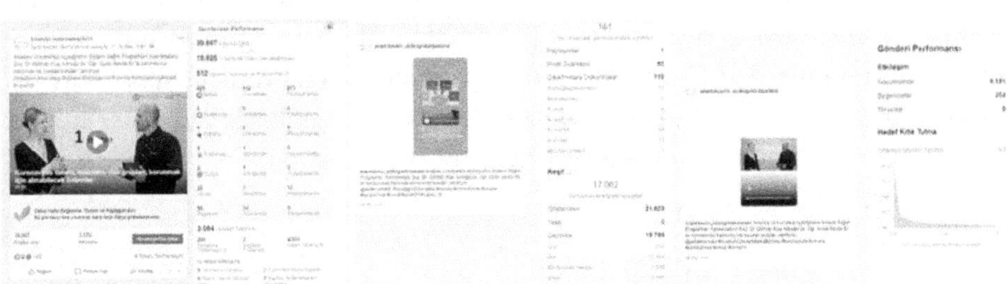

MASSIVE OPEN ONLINE COURSES

Within the scope of "social responsibility understanding based on the principles of giving back what is received from the society and upholding social benefits in corporate activities", Anadolu University introduces many social projects. The knowledge and experience of the Open Education system of Anadolu University are very important in terms of fulfilling the gaps in the education of the society. The format which was first started as a result of the openness policy in education in the Open Education System of Anadolu university as "Massive Open Online Courses" in 2008 was transformed into an effective learning area known as "Akadema" in 2015. With the infrastructure and opportunities it offers, Akadema is an interactive learning environment in which many participants can get involved at the same time. Through Akadema platform, various courses and videos, applications, assignments, tests etc. on a variety of topics are prepared by expert academics in their fields and offered to the participants. Participants can actively take part in the education through discussion platforms offered. In this sense, Akadema is a massive and free learning environment open to everyone. On this platform, it is possible to determine how long the participant has stayed in the system, to what extent s/he followed the subjects, whether s/he has completed the projects and assignments, and success rate in the tests (Akadema, 2019).

Through Akadema, which is a Massive Open Online Course (MOOC) platform, everyone can have access to basic information on first aid and coronavirus free of charge. The titles of the courses offered directly on health issues are:

a) Basic First Aid Knowledge
b) Coronavirus

AKADEMA: BASIC FIRST AID KNOWLEDGE

In 7 semesters, approximately 5,000 students registered to the Basic First Aid Course, which started to be offered in November 2016. Owing to the importance attached to the First Aid Course on Akadema, course achievement score was set between 75-85. As a result of education and practices, feedback and tests, around 400 students were entitled to get a certificate of participation. Four of the first aid subjects were offered on Akadema:

1. Basic First Aid Knowledge
2. Basic life support in adults
3. Basic life support in children and babies, and first aid in airway obstructions

4. First aid in bleedings and chest injuries

Related to the headings mentioned above, general theoretical explanations, video recordings, discussion forums/question-answer sessions, assignments given as end-of-course project and feedback, weekly test analyses were shared with the participants.

Figure 7. Akadema First Aid Course Demo Video

Videos and PowerPoint presentations on the subject were weekly shown to the participants for four weeks. Feedback on what they should do related to the subjects was received. Questions and answers related to the subjects were discussed and shared in the forums.

Figure 8. A scene from First Aid Course on Akadema

DISCUSSION AND CONCLUSION

When we consider historical process starting from the day humans were created until today, we can see many important events. In fact, some of these events started and ended epochs. It may have started or ended an epoch, but the invention of the Internet and its widespread use in today's information and technology age is significant. When the Internet was invented, it was initially used in our society mainly for conversation, surfing environment and messaging. However, in time, especially in recent years, with the development of new communication technologies in the Internet and Internet based social media environments, these environments started to be used as development and learning tools. It was not only the public institutions and organizations that used these environments for these purposes, but many independent and private organizations and companies also used this environment for similar purposes. As development and learning processes were involved, institutions in the area of education quickly participated in this process.

Structures which were able to meet the needs and desires of the society which developed as a result of fast and continuous changes in technology and easy access of societies to information and orientation towards another dimension in terms of socialization and learning have managed to survive in this process. Those who could not adapt and develop themselves were left behind. Now, public institutions and organizations and independent and private companies started to reach people they provide services through new communication technologies instead of traditional mass communication tools. Thanks to these environments, they have been able to serve their clients faster and more effectively. Since new communication technologies provide people with an environment in which they can share and discuss their ideas and thought sat the comfort of their private homes, the feedback is also provided over these platforms (Solmaz, 2013). New communication technologies offer many significant advantages such as being interactive, low cost, access to more people and instant interaction (Bakan, 2008). Thus, those who noticed these advantages began to apply these new communication technologies to education field. Anadolu University and Anadolu University Open Education system also makes use of new communication technologies, especially in massive education field. The Open Education system, which has almost a million students, offers massive education in a variety of areas through new communication technologies. It employs the Internet and social media on its massive open education platform, Akadema.

The subjects which are determined to be lacking regarding the society are identified, and educational materials prepared on these subjects are offered to the users free of charge. Also, in line with the suggestions, criticisms and requests from the users, education on other subjects are provided. Among these subjects are Basic Health Knowledge, First Aid, and Coronavirus. Educational materials prepared on these subjects are provided to the users in the form of "Explain in 90 Seconds" videos on Facebook. Similarly, these subjects are also opened to the use of the participants on Akadema platform. On both platforms, the subjects had a significant rate of follows, shares and likes. Positive responses received from the users also point to the importance of these videos offered in a short time. What is important here is to inform people in the correct way on the issue that would cause public unrest. Information sharing on coronavirus pandemic and Akadema courses have gained importance in this context. They prevented disinformation that would cause panic in people.

Face-to-face education in the form of courses and in-service training on basic health, first aid and coronavirus can be provided. However, there are three important points that should be considered here:

1. Given that not everyone could afford these trainings, such training should be provided in "environments where everyone can heave access free of charge."

2. It should also be considered that the gathering of people together in face-to-face education environments can increase the spread of pandemics such as coronavirus.
3. Since it is known that students prefer video presentations rather that face-to-face education, education aimed at society should be moved to platforms where individuals can follow at any time (Traphagan, Kucsera, Kishi, 2010).

Educational materials (presentation, video, demonstration, application etc.) to be prepared on the basis of these three important factors should be provided in social media environments so that everyone can have access to them. Thus, people will obtain correct and effective information from reliable sources and can form ideas on a variety of subjects.

REFERENCES

Akadema. (2019). *Anadolu Üniversitesi Kitlesel Açık Çevrimiçi Ders Platformu*. http://akadema.anadolu.edu.tr

Akbaş, O., & Özdemir, S. M. (2002). Avrupa Birliğinde yaşam boyu öğrenme. *Milli Eğitim Dergisi, 155*(156), 112–126.

Aktan, C. C., & Börü, D. (2007). Kurumsal sosyal sorumluluk. *Kurumsal sosyal sorumluluk: İşletmeler ve sosyal sorumluluk*, 11-36.

Arklan, Ü. (2016). Sosyal medyanın siyasal amaçlı kullanımı: Ağ kuşağının kullanım alışkanlıkları üzerine bir araştırma. *Gümüşhane Üniversitesi İletişim Fakültesi Elektronik Dergisi, 4*(2).

Aydin, A. F. (2017). Sosyal medya ve kurumsal markalaşma ilişkisi: anaakım iletişim kuramları perspektifinden genel bakış. *İletişim Kuram ve Araştırma Dergisi*, (44).

Billman, G. E. (2020). Homeostasis: The Underappreciated and Far Too Often Ignored Central Organizing Principle of Physiology. *Frontiers in Physiology, 11*, 200. doi:10.3389/fphys.2020.00200 PMID:32210840

Budak, Y. (2009). Yaşamboyu öğrenme ve ilköğretim programlarının hedeflemesi gereken insan tipi. *Gazi Üniversitesi Gazi Eğitim Fakültesi Dergisi, 29*(3).

Cassin, R. (2002). La formation Tout au Long de la Vie, Séminaire La formation professionnelle. Ecole Nationale d'Administration Paris Comission Européenne. Union Europeenne. Education et Formation Tout au Longue de la Vie. Education et Formation Jeunesse.

da Silva, J. L., de Oliveira, R. F., Bertin, L. D., de Oliveira, P. D., Pires, F. B., Spadaccini, A. F., ... & de Almeida Pires-Oliveira, D. A. (2018). Physical inactivity and associated factors in adults of a basic health unit: An observational cross-sectional study. *Manual Therapy, Posturology & Rehabilitation Journal, 16.*

Duman, T. (2001). Avrupa Birliği Eğitim Programları 'Leonardo da Vinci Programı'. *Milli Eğitim Dergisi*, (150).

FREREF. (2004). *Lifelong Learning The Challenges*. Printed by l'Imprimerie Ecully Graphic.

Göksan, T. S., Uzundurukan, S., & Keskin, S. N. (2009). Yaşam boyu öğrenme ve avrupa birliği'nin yaşam boyu öğrenme programlari. *İnşaat Mühendisliği Eğitimi Sempozyumu, 143.*

Kozikoğlu, İ., & Altunova, N. (2018). Öğretmen Adaylarının 21. Yüzyıl Becerilerine İlişkin Öz-yeterlik Algılarının Yaşam Boyu Öğrenme Eğilimlerini Yordama Gücü. *Journal of Higher Education & Science/Yüksekögretim ve Bilim Dergisi, 12*(3).

Kuş, G. (2018). Doğal Afetler Sonrasinda Insanlarin Ilkyardim Bilgi Ve Beceri Düzeyi: Ne Yapabiliriz? *Anadolu University of Sciences & Technology-B: Theoretical Sciences, 6*(2).

Lee, M., & Friedrich, T. (2011). Continuously reaffirmed, subtly accommodated, obviously missing and fallaciously critiqued: Ideologies in UNESCO's lifelong learning policy. *International Journal of Lifelong Education*, *30*(2), 151–169. doi: 10.1080/02601370.2010.547619

Lengrand, P. (1989). Lifelong education: growth of the concept. In *Lifelong Education for Adults* (pp. 5–9). Pergamon. doi:10.1016/B978-0-08-030851-7.50008-4

Özdemir Çakır, H. (2009). *Kurumsal sosyal sorumluluğun marka imajına etkisi.* Academic Press.

Özmusul, M. (2018). Eğitim Ve Yetiştirme 2020 Stratejisi: Hedef Göstergelere Göre Ülkelerin Durumu. *Journal of European Education, 2*(1).

Öztürk, Y. E., & Kıraç, R. (n.d.). Sağlık ve Hastalık. *Scientific Developments*, 382.

Rodriguez, J. A. O., Ladd, M., & Brandis, D. (2020). Abdominal Thrust Maneuver. In *StatPearls*. StatPearls Publishing.

Sağlam, M., Özüdoğru, O. F., & Çiray, F. (2011). Avrupa Birliği eğitim politikaları ve Türk Eğitim Sistemi'ne etkileri. *Yüzüncü Yıl Üniversitesi Eğitim Fakültesi Dergisi, 8*(1), 87–109.

Scott, J. (2011). Social network analysis: Developments, advances, and prospects. *Social Network Analysis and Mining*, *1*(1), 21–26. doi:10.100713278-010-0012-6

Til, U. D. A. (2020). Yeni Koronavirüs hastalığı Hakkında bilinmesi gerekenler. *Ayrıntı Dergisi, 8*(85).

Toprak, M., & Erdogan, A. (2012). Yaşamboyu Öğrenme: Kavram, Politika, Araçlar ve Uygulama [Lifelong Learning: Concept, Policy, Instruments and Implementation]. *Journal of Higher Education and Science*, *2*(2), 69. doi:10.5961/jhes.2012.036

Traphagan, T., Kucsera, J. V., & Kishi, K. (2010). Impact of class lecture webcasting on attendance and learning. *Educational Technology Research and Development*, *58*(1), 19–37. doi:10.100711423-009-9128-7

Ural, Y. Ç. (2017). Solunum, kalp ve dolaşım sistemiyle ilgili acil durumlar ve ilk yardım. *Pegem Atıf İndeksi*, 110-142.

Usta, İ., Uğur, S., Okur, M. R., & Kuş, G. (2019). A Study on The Use of Social Media of Open Education System Students: The Example of AofAnadolum. *Kastamonu Education Journal, 27*(2).

Wain, K. (1993). Lifelong education: Illiberal and repressive? *Educational Philosophy and Theory*, *25*(1), 58–70. doi:10.1111/j.1469-5812.1993.tb00186.x

Wain, K. (2001). Lifelong learning: Small adjustment or paradigm shift? In *International handbook of lifelong learning* (pp. 183–198). Springer. doi:10.1007/978-94-010-0916-4_12

Zu, Z. Y., Jiang, M. D., Xu, P. P., Chen, W., Ni, Q. Q., Lu, G. M., & Zhang, L. J. (2020). Coronavirus disease 2019 (COVID-19): A perspective from China. *Radiology*, *296*(2), 200490. doi:10.1148/radiol.2020200490 PMID:32083985

Chapter 6
Measuring Student Satisfaction Level Regarding Aesthetic and Instructor Dimensions of Web-Based Distance Learning Programs:
The Case of Spiritual Guidance Training Program

Serhat Koca
Anadolu University, Turkey

Murat Aytaş
https://orcid.org/0000-0003-2744-0519
Selcuk University, Turkey

ABSTRACT

Today, many educational institutions provide their content to their recipients through web-based distance learning applications. While providing this education method to the recipients through communication technologies, the aesthetic design and the instructor dimension of the content are of utter importance. In this framework, the study measured the level of student satisfaction regarding the aesthetic and instructor dimension in web-based distance learning programs within the case of the Spiritual Guidance program. Accordingly, the study of distance learning programs in terms of aesthetic and instructive dimensions will contribute to the research that will be conducted in this direction.

DOI: 10.4018/978-1-7998-4534-8.ch006

INTRODUCTION

Technology played a decisive role in the development and progress of societies. In the historical development process, humans first imitated nature and developed various tools. Thanks to the culture they created, the production of these tools passed on to the next generation. This cultural transfer continued through education, and education has played a vital role in shaping the future of societies with several approaches and methods in the historical development process of humanity. In this context, the relationship between education and technology has developed each other, and particularly in the present century, education and technology interacted with each other very quickly and gained an almost integrated view. With the invention of the printing press, the rapid reproduction of printed resources, the advances in transportation and communication technologies, and the spread of printed resources have paved the way for the phenomenon of distance education, which has increased the interaction between education and society in an unprecedented manner. The new process has started with the widespread use of the internet in distance education approaches, which have gained popularity with the development of technological tools that provide flexibility for the individual learning environment such as video cassette, CD, DVD, etc.

Especially in this century, developments in information and communication technologies have brought about important changes in the field of education, as in all areas of life. In today's information society, where information has become the determinant of economic production, information production, intellectual capital, and qualified human factor have gained importance. In line with this, the use of internet technology in education has become an inevitable necessity in both traditional education and distance education, as it stands out for the continuity of education. Providing continuity of education and the increasing need for information has increased the importance and prevalence of distance education while the new methods in distance education enabled the widespread usage of the opportunities provided by new technologies. Undoubtedly, Web-Based Distance Learning application is the most common distance learning method that utilizes the latest information and communication technologies.

Another field where web-based distant education applications is lifelong learning programs. Lifelong learning involves all sorts of formal and distant education activities that will contribute to the vocational and individual development of the individuals of all ages. Especially, web-based distant education applications enable the participants in receiving education on numerous subjects notwithstanding the limits of space through improving the opportunities in terms of lifelong learning. In Turkey, one of the significant and modern programs which contribute to lifelong learning through distant education is Spiritual Guidance program. Spiritual guidance is a distant education program which was prepared in order to satisfy the needs of consultants with religious and moral orientation and considering the issues which society needs in this field. The program which was put into practice open to the participation of all the individuals in terms of providing contributions to the both individual and vocational development of the individuals who work in this field became very popular in Turkey. The content production, registry and coordination of spiritual guidance programs are conducted by lifelong learning centers of the universities. The Spiritual Guidance distant education program which is conducted by continuing education center of Selçuk University reached to numerous students all over the country as the first program in this field.

Various educational institutions in Turkey transmit their contents to their students of all ages through web based distant education applications in terms of lifelong learning. Evaluating the satisfaction level of the course contents among the students throughout the process of applying distant education programs and gains significance in terms of revealing the sufficiency of suchlike programs. There are only a few studies which satisfaction levels of the aforementioned distant education programs are evaluated in terms of aesthetical design and deductive dimension of those programs. The basic problem of this study is to research the way students evaluate the aesthetical design and deductive dimension of the contents of distant learning related to lifelong learning considering Spiritual Guidance program.

In this framework, the study measured the level of student satisfaction regarding the aesthetic and instructor dimension in web-based distance learning programs within the case of the Spiritual Guidance program. Accordingly, the study of distance learning programs in terms of aesthetic and instructive dimensions will contribute to the research that will be conducted in this direction.

CONCEPTUAL FRAMEWORK

In this section, the relationship between web-based distance learning programs and the aesthetic/instructor dimension and satisfaction in the evaluation of these programs is explained in a theoretical framework.

Web-Based Distance Learning Programs

Communication technologies have become effectively useful in all areas of life, with the new social formation, which Daniel Bell defines as the post-industrial society and Spanish sociologist Manuel Castells describes as the "network society" and is now referred to as the "information society" (Bell, 1973; Castells, 2005). As the developments in communication technologies are also used in the field of educational technologies, individuals of the modern age can access knowledge in any field they want through distance learning as a part of everyday life whenever they want. Compared to formal education, perhaps the most advantageous aspect of distance education is that it can reach students or participants with relevant content at any time and place, regardless of time and place. Along with the distance learning method, the instructor and students can carry out their educational processes by choosing one of two different types: synchronous (simultaneous) and asynchronous (not simultaneous) (Akbulut, 2017: 2019). There are quite different applications that involve different relationships and interactions in terms of the method of learning/teaching through distance education. Web-Based Distance Education (WBDL) application is one of the most preferred of these methods. The field of communication technologies, where the Internet is the infrastructure, has advanced by first launching the concept of WBDL and then its various applications. Web-Based Distance Learning is an electronic learning method that is carried out independently from time and place and enables people to get training (Akca and Aydogan, 2014: 178). As in the world, many universities and educational institutions carry out various training and certificate programs through web-based distance learning platforms in Turkey. WBDL is not only a place for users to train individually, but also for dealing with interactions between the course instructor or other individuals taking the course, such as messaging, asking questions, forums, and e-mail. The most important difference between the traditional method and the WBDL system is that instructors and students do not have to be together at the same time and place (Cakmak, 2013: 266). Web-based learning platforms offer students almost unlimited access to video content, e-learning software, multimedia resources, online learning hyperlinks, online communication forums, and presentations (Hsu, 2017: 20). Gulnar (2008: 262) stated the common features in the definitions about distance education as follows:

- Distance learning systems follow an effective method of accessing resources of knowledge and providing the students with these resources.
- Distance learning systems utilize technology at the highest level in the execution of educational services.
- In the distance learning process, students and instructors are not bound by time and place.

WBDL applications have facilitated access to information by making classroom teaching and learning available at any time, both through computers and smartphones.

Aesthetic and Instructor Dimensions of Distance Learning Programs

Web-based learning systems are at different levels in terms of having teaching and design practices that can affect the quality of learning (Alomyan, 2017: 191). There are many criteria in designing and evaluating a web-based teaching environment. Within the scope of the study, evaluation criteria are outlined under four major aspects: educational design, technical dimension, aesthetic dimension, and instructor dimension (Gulnar, 2008, Duffy, 2008; Crismond and Adams 2012; Sahin and Yildirim, 1999; Yang, Guo and Yu, 2016, Nichols, 2016, Lee, 2018).

The aesthetic dimension of distance learning programs consists of many elements ranging from the technical competence of the course materials (video, animation, graphics, tables, etc.) to the graphic design of the visual material to be used. Visual materials, which provide concrete insight in learning activities, are an important reinforcer and reminder, as well as a crucial motivation tool for the learning process and increasing the focus on course content. According to the findings of the Cone of Experience suggested by Edgar Dale, 1% of what we learned is through tasting experience, 1.5% with touching experience, 3.5% thorough smelling experience, 11% through hearing experience, 83% through seeing experiences (Lee and Reeves, 2007: 56). Therefore in the web-based distance learning programs, the aesthetic dimension should be simple, understandable, and compatible with the goals and objectives of the course. The aesthetic dimension is quite important as the digital literacy level and ways of interacting with web-based material will differ among individuals having personal differences in web-based distance learning programs. While screen-based learning experience that teaching designers emphasize today provides more strategic possibilities, as well as, providing both educational and institutional conveniences (Nichols, 2016; 40). Learning design and content elements that create a learning ecology in web-based distance learning programs should be interconnected and dynamic (Duffy, 2008: 127).

One of the most important components of web-based distance learning programs is the instructor. The instructor and the student, who are physically located in separate places, communicate and interact in web-based distance learning environment through the Internet. One of the biggest advantages of web-based distance learning for students is that it allows them to learn whenever and wherever they want. The flexible learning model offered by web-based distance learning not only requires learners to motivate themselves but also differentiates the roles and responsibilities of the teacher from traditional education. The general model of distance learning suggests combining the following three methods in learning (Ben-Chayim and Offir, 2019).

1. The expert instructor remotely performs training synchronously.
2. The lessons recorded by the expert instructor are delivered asynchronously.
3. Asynchronous distance learning is conducted via a website where the course is carried out.

In web-based distance learning programs, it is not necessary for the teacher to master the content of the subject, to use a correct, fluent language, to keep the interest of students intact throughout the narration, and to support the lecture with current examples, as it is in traditional face-to-face education. Also, during web-based distance learning, the instructor should be able to provide a high level of interaction between himself and those attending the course and increase motivation for the attendance and continuity of the learning process. The research conducted by Jones and Rodriguez (2010) revealed that the teacher's competence in web-based distance learning was an important determinant considering the factors increasing the students' satisfaction in both synchronous and asynchronous lessons are examined. In WBDL, the instructor's poor competence in distance learning will prevent effective learning as well as decreasing the motivation of the learners. In the study of Kaysi and Aydemir (2017), it was revealed that the instructor's use of effective communication management to attract the attention of the participants and keeping the interaction intact throughout the course was a factor that increased the level of learning.

Distance Learning and Satisfaction

Although the concept of satisfaction has several features that have different meanings for each individual, it can be generally defined as the rating of meeting the expectations, emotions, and suitability level of the environment or situation (Korkmaz, Cakir, and Tan, 2015). Some many elements and variables affect these satisfaction levels in terms of user satisfaction with content transmitted through distance learning. Considering the studies on satisfaction in distance learning, there are many dimensions such as teaching, ease of use, flexibility, technological comfort, group work, interaction, course design and aesthetics, and the teaching dimension (Virtanen et al., 2017). In the web-based distance learning method, system interface affects user satisfaction in factors such as content design and system functionality (Cheng, F et al., 2017).

Many studies in the literature demonstrate the high satisfaction with learning through web-based distance learning programs (Korkmaz, Cakir, and Tan, 2015; Klein, Cidral, W. A et al. 2018). According to the study of Zhang et al. (2006), the learning performance and satisfaction level of students who participate in distance learning programs with interactive video are much higher than in other groups.

In the study of Korkmaz, Cakir, and Tan (2015) on university students, the students state that they are very satisfied with the learning processes offered to them through the system. Also, the students' perception of satisfaction with the electronic content used in the distance learning environment and the interaction and evaluation processes related to these contents is quite high. In the study conducted by Metin, Karaman, and Sastim (2017) with 216 students on the evaluation of the efficiency of the English course given by distance learning methods, 57% of the students who took distance learning courses in a virtual environment stated that they did not consider these courses as important as face-to-face courses, 67% did not find distance learning courses as efficient, 78% did not find the process of courses interesting and enjoyable, and 75% stated that they were more successful in face-to-face courses.

One of the studies on student satisfaction in web-based platforms is the study of Tsu (2017) with 302 students. According to the results of the satisfaction study conducted by the instructors who have diversified the learning process by using the web-based platform in English language teaching, it has been revealed that there is a positive correlation between the inclusion of web-based learning materials and learning satisfaction and interest (Tsu, 2017: 20). One of the studies examining the factors for the determination of success in web-based learning ecosystems is Cidral's (2018) study with 301 online students. According to the results of the study, there is a positive correlation between the use of web-based learning ecosystems and the quality of cooperation and the success factor.

Methodology

The study has a "descriptive" research pattern that aims to measure the satisfaction level of students enrolled in web-based learning programs. From this point of view, the study also has a "relational" design dimension based on the idea that the descriptive findings obtained will differ according to the individual qualities of the participants. To achieve these goals, descriptive analyses (frequency analysis, central tendency statistics), one-way analysis of variance, independent sample t-test, and correlation analysis were conducted.

This study aims to measure the satisfaction levels of the participants who enrolled in the Spiritual Guidance distance learning program in terms of their aesthetic and instructor dimensions. In line with this purpose, the answers to the following research questions were sought:

1. What are the satisfaction levels in the overall evaluation of the spiritual guidance program?
2. What are the levels of satisfaction in the sub-dimensions?

3. Do the levels of satisfaction differ according to the individual characteristics of the participants? Application of the Research and Sampling

The study was carried out between 6 January 2020 and 10 February 2020 with 259 students who received Spiritual Guidance Training through distance learning from Selcuk University Continuing Education Center through the online survey method. Since it was possible to reach the entire study universe, no sampling has been performed.

Table 1. Distribution of Individual Differences of Participants

	Group	Frequency	Percentage
Age	18-25	63	25.2
	26-33	58	23.2
	34-41	78	31.2
	42 +	51	20.4
Educational Background	High School	13	5.3
	Associate Degree	87	35.8
	Bachelor's Degree	122	50.2
	Postgraduate	24	9.8
Employment Status	Public Sector	138	57.2
	Private Sector	20	8.3
	Self-Employed	8	3.3
	Unemployed	75	315
Gender	Male	163	63.6
	Female	94	36.7

Measurement Tool

Based on the literature review and the data they obtained from previous studies, the researchers developed a distance learning assessment scale consisting of 30 questions in two dimensions. The study in which the survey was used as a data collection tool used a 5-point Likert scale in which 1 represents the lowest level and 5 represents the highest level of satisfaction. The researchers found the reliability value (alpha) of the scale they developed as 0.928.

In this study, alpha reliability values were calculated separately for the entire scale and sub-dimensions. The alpha reliability value for the entire scale was 0.928, while the sub-dimensions were 0.821 for the aesthetic evaluation dimension and 0.922 for the instructorevaluation dimension. The significant positive correlations between the spiritual guidance training program general evaluation index and the sub-dimensions of aesthetic evaluation ($r = .928$, $p < .001$) and instructor dimension ($r = .749$, $p < .749$, $p < .001$) reveal the structural validity of the scale.

Table 2. Reliability Values of the Survey

Variables	*Number of Items*	*Cronbach's Alpha*
Aesthetic Evaluation	16	.821
Evaluation of the Instructor	14	.922
General Evaluation of the Program	30	**.928**

RESULTS

Descriptive statistics, regression analysis, correlation analysis, and one-way analysis of variance were performed to obtain descriptive and relational findings.

Results of Central Tendency Analysis Regarding General Evaluation of Spiritual Guidance Program

Descriptive analyses were performed to reveal the level of satisfaction that the participants gained regarding the distance learning program and their instructor dimensions. In the first stage, central tendency statistics were applied to a total of 30 items that measure satisfaction levels associated with the aesthetic and instructor dimension in Spiritual Guidance Training. As can be seen in Table 3, the related items are listed from high to low according to their arithmetic mean values. Accordingly, the first five items with the highest satisfaction level of those who enrolled in the spiritual guidance distance education program are listed as follows:

1. General satisfaction level from the instructor qualifications of the distance learning program (Instructor Dimension)
2. The instructor's outfit who conduct the lectures do not distract me from the lesson (Instructor Dimension)
3. Level of harmony and consistency between different screens of the system (Aesthetic Dimension)
4. The level of simplicity in the overall visual design of the system (Aesthetic Dimension)

5. The level of fluency and use of the proper language of the teacher who provides the lecture presentations (Aesthetic Dimension)

On the other hand, the aesthetic and instructive dimensions that the students are satisfied at the lowest level are listed as follows:

1. The duration of the courses (live-recording) is short (Aesthetic Dimension)
2. The level of knowledge presented in the visual design used in the courses (Aesthetic Dimension)
3. The duration of the courses (live-recording) is long (Aesthetic Dimension)
4. Level of keeping the interest intact during the lecture of the instructor (I.D) Dimension)
5. The level of interaction between the instructor who gives the lecture and the student (Instructor Dimension)

According to the results in Table 4 (total), the level of satisfaction ($\chi = 3.75$) related to the general aesthetic and instructor dimension of the participants enrolled in the Spiritual Guidance distance learning program is "high". Examining the dimensions that made up the satisfaction related to the distance learning program, the instructor dimension ($\chi = 4.19$), which was related to general educational satisfaction, was found to be the item with the highest level of satisfaction ($\chi = 4.19$).

Analyzing the average values of the first three items in the satisfaction ranking, all three correspond to the "high" level of satisfaction, which is in the range between 3.41-4.20 on the 5-point scale.

On the other hand, the least satisfied dimension was in the form of aesthetic dimension regarding the short recording times of the courses ($\chi = 2.5$) and the clearness of the content provided in visual design ($\chi = 3.1$). The dimensions with the lowest satisfaction correspond to the "medium" level of satisfaction on the 5-point scale. In other words, even the aesthetic and instructor dimensions, which have the least satisfaction levels, are not below the medium level. This value means that distance learning program participants' satisfaction levels in aesthetic and instructive dimensions are "high" (five-point Likert scale was used to measure the aesthetic and instructive dimension satisfaction level).

Considering the evaluations of the participants regarding the Spiritual Guidance program in terms of sub-dimensions according to the results in Table 5, the average of the overall evaluation of the aesthetic dimension was 3.71 and the overall evaluation of the instructor dimension was 3.81.

The answers naturally range from 1 to 5, therefore, a range of 0.80 (4/5) was used to determine the level: 1.00-1.80 = very low; 1.81-2.60 = low; 2.61-3.40 = medium; 3.41-4.20 = high; 4.21-5.00 = very high.

Table 3. Statistical Results of the Central Tendency Regarding General Evaluation of Spiritual Guidance Program

	N	Minimum	Maximum	Mean	S. Deviation
Level of general satisfaction from the educational qualifications of the distance learning program	259	1.00	5.00	4.1969	1.00572
The instructor's outfit who conduct the lectures does not distract me from the lesson.	254	1.00	5.00	4.0984	.86782
Level of harmony and consistency between different screens of the system	255	2.00	5.00	4.0863	.65839
The overall visual design of the system is not complicated	258	1.00	5.00	4.0853	.82312
The instructor who provides the lecture presentations uses fluent and proper language	254	1.00	5.00	4.0787	.81591
The technical quality level of course materials such as synchronous (live), recording, video, audio, etc.	258	1.00	5.00	4.0698	.94343
The instructor who teaches the subject is knowledgeable about the course content	256	1.00	5.00	4.0195	.93913
Level of overall satisfaction regarding the visual dimension (aesthetic) of the distance learning program	259	1.00	5.00	4.0116	1.02103
The instructor who makes the lecture gives integrity in course	256	1.00	5.00	4.0000	.90315
The instructor who gives the lecture provides current examples about the subject	255	1.00	5.00	3.9176	.87230
The graphic design of the course materials in the system is complete and consistent.	257	1.00	5.00	3.9066	.84724
There is no technical errors and failures in the recording videos of the course presentations	256	1.00	5.00	3.8867	.97762
The visual atmosphere (light, color, camera angle/scales, etc.) established in the lecture presentations facilitates focusing on the content	256	1.00	5.00	3.8789	.96862
The graphic elements used in the lecture presentations are sufficiently sized.	255	1.00	5.00	3.8275	.93153
Interactive items in recording video presentations are sufficient.	256	1.00	5.00	3.7695	1.00079
Course materials such as synchronous (live), recording, video, audio, etc. are arranged to intensify the student's attention	255	1.00	5.00	3.7490	1.08304
Graphics and animation used in the program are suitable for the content of the courses	258	1.00	5.00	3.7481	1.01855
I found the presentation of the instructor who performed the lecturing effective	254	1.00	5.00	3.7126	1.05579
The instructor, who teaches the subject, explains the content from concrete to abstract.	254	1.00	5.00	3.7087	.84451
The presentation of the instructor, who gave the lecture, increased my motivation to research on the subject	253	1.00	5.00	3.6996	1.03351
Visual materials are not monotone.	253	1.00	5.00	3.6640	1.03595
The presentation of the instructor who gives the lectures increases my motivation for the lesson	253	1.00	5.00	3.6640	1.00878
The visual design of the lessons intensifies the student's attention and is aesthetic.	257	1.00	5.00	3.6498	1.11888
The tone of voice and intonation of the teacher who gives the lectures increases my interest in the lesson	255	1.00	5.00	3.6471	1.08721
The use of body language of the instructor who teaches the subject supports the lecture	255	1.00	5.00	3.5451	1.06341
The instructor who gives the lectures attaches importance to the interaction with the student	254	1.00	5.00	3.5394	1.08367
The instructor who makes the lecture can keep the interest intact during the lecture	255	1.00	5.00	3.5020	1.13594
The duration of the courses (live-recording) is long	256	1.00	5.00	3.3320	1.26612
The content presented in the visual design used in the lessons is clear and understandable	257	1.00	5.00	3.1712	1.38979
The duration of the courses (live-recording) is short.	255	1.00	5.00	2.5098	1.11843

Table 4. Results of the Descriptive Analysis Regarding the General Satisfaction Level with the Training Program

	N	Minimum	Maximum	Mean	S. Deviation
General Evaluation of Distance Learning Program According to Aesthetic and Instructor Dimensions	259	1.93	5.00	3.7592	.57109

Table 5. Results of Descriptive Analysis Regarding Aesthetics and Instructor Dimension Satisfaction

	N	Minimum	Maximum	Mean	S. Deviation
General Evaluation of Aesthetic Dimension	259	2.13	5.00	3.7117	.53766
General Evaluation of the Instructor Dimension	259	1.64	5.00	3.8191	.69582

Evaluation of Spiritual Guidance Program and Individual Characteristics

This section examined whether the satisfaction levels of the participants regarding the Spiritual Guidance program differ according to their individual characteristics. In this sense, the first examination was carried out based on the age.

Evaluation of the Overall Satisfaction and Age

A significant difference between the ages of the participants and the aesthetic evaluation ($r = -.050$, $p > .05$), instructor dimension ($r = .024$, $p > .05$) and the general evaluation index ($r = -.013$, $p > .05$) was not found.

Table 6. Correlation Between the Age and Results of the Satisfaction Level according to Correlation Analysis (Pearson r)

	Aesthetic Dimension	Instructor Dimension	General Evaluation
Age	-.050	.024	-.013

Evaluation of Overall Satisfaction and Gender

According to the results in Table 7, the average of evaluation of overall satisfaction of the male participants ($x = 3.90$) is significantly higher than female participants (x = 3.67) (t = -3.119, p <.01). The appreciation level of male participants is higher than female participants in terms of both aesthetic and instructor dimensions and overall evaluation.

Table 7. Results of Independent Sample T-Test for Satisfaction Level and Gender

Satisfaction Level and Dimensions	Gender	N	Mean	t-Test	Sig
Aesthetic Dimension	Female	163	3.63	-2.876	.004
	Male	96	3.83		
Instructor Dimension	Female	163	3.71	-3.059	.002
	Male	96	3.98		
General Evaluation	Female	163	3.67	-3.119	.002
	Male	96	3.90		

Evaluation of Overall Satisfaction and Training

One-way analysis of variance (ANOVA) was applied to test the prediction that the participants' satisfaction levels of the Spiritual Guidance distance learning program would differ according to their education levels. According to the results in Table 9, the participants' satisfaction levels of the relevant program do not differ significantly according to their education levels (F = .982, p>.05) Also, the level of satisfaction from the sub-dimensions of the program does not differ significantly according to the education level (F = 1.433, p>.05; F= .793, p>.05).

Evaluation of Overall Satisfaction and Employment Status

One-way analysis of variance (ANOVA) was applied to test the prediction that the participants' satisfaction levels of the Spiritual Guidance distance learning program would differ according to their employment status. According to the results in Table 10, the participants' satisfaction levels of the relevant program do not differ significantly according to their employment status (F = .670, p>.05) Also, the level of satisfaction with the sub-dimensions of the program does not differ significantly according to the employment status (F = 1.917, p>.05; F= .151, p>.05).

Table 8. Results of ANOVA for Satisfaction Level and Education Level

Satisfaction Level and Dimensions	Education	N	Mean	F-Test	Sig
Aesthetic Dimension	High School	13	3.65	1.433	.499
	Associate Degree	86	3.75		
	Bachelor's Degree	117	3.64		
	Postgraduate	43	3.82		
Instructor Dimension	High School	13	3.81	.793	.234
	Associate Degree	86	3.91		
	Bachelor's Degree	117	3.75		
	Postgraduate	43	3.80		
General Evaluation	High School	13	3.72	.982	.402
	Associate Degree	86	3.82		
	Bachelor's Degree	117	3.69		
	Postgraduate	43	3.80		

Table 9. Results of ANOVA for Satisfaction Level and Employment Status

Satisfaction Level and Dimensions	Employment Status	N	Mean	F-Test	Sig
Aesthetic Dimension	Public Sector	135	3.66	1.917	.127
	Private Sector	20	3.89		
	Self-Employed	11	3.50		
	Unemployed	93	3.75		
Instructor Dimension	Public Sector	135	3.81	.151	.929
	Private Sector	20	3.91		
	Self-Employed	11	3.82		
	Unemployed	93	3.80		
General Evaluation	Public Sector	135	3.73	.670	.571
	Private Sector	20	3.90		
	Self-Employed	11	3.65		
	Unemployed	93	3.77		

CONCLUSION AND DISCUSSION

There are numerous applications within the method of learning/teaching through distant education which include various relationships and interactions. Among the aforementioned methods, one of the most frequently preferred ones is Web Based Distant Education (WBDE) program. The communication technologies which internet creates the infrastructure displayed a progress in the field of education through initiating the term of WBDE initially and its numerous applications laterally. The Web Based Distant Education which is frequently employed within the lifelong learning programs nowadays is an electronic learning program which is conducted independently from time and space in the simplest terms and enables individuals receive education.

In this study which employs the "descriptive" research design, the satisfaction level among the students was evaluated in terms of aesthetical and deductive dimension of the Spiritual Guidance program which is a WBDE application that internet creates the infrastructure. In the general evaluations of the sampling which consists of the participants of Spiritual Guidance Education Program, it was found that the satisfaction levels were "high" in terms of aesthetical dimension ($\chi = 3,7$) and deductive dimension ($\chi = 3,8$). The lowest levels of satisfaction of the participants for the aesthetical dimension are the short ($\chi = 2,5$) and long ($\chi = 3,3$) durations of the lesson and information level ($\chi = 3,1$) presented in the visual designing which is employed in lessons. In terms of deductive dimension, the lowest levels of satisfaction among the participants were the level which teacher keeps the interest alive ($\chi = 3,5$) and the level of interaction with the students ($\chi = 3,5$).

The results obtained in the conducted study displays from the perspective of aesthetical dimension that the duration of the courses should be determined regarding the cognitive strategies of the individuals so that their individual learning needs can be satisfied. The low levels of satisfaction among the participants due to short or long duration of the lessons reveal the necessity for diversifying the learning materials in terms of duration. The information level of the visual educational materials which the participants had low levels of satisfaction is an evidence of the need for visual materials which increase the attention to the contents of the course and reinforce them. Visual materials which are important materializers and are remindful in learning process should be rich in terms of information level and should be supported with interaction.

As stated in the literature, interaction is an important factor in the relationship between the teacher and the learner which increases the levels of learning and satisfaction. When the results of this study is evaluated in terms of deductive dimension, it was found that the interaction was at insufficient levels in the Spiritual Guidance Education Program and the teachers should keep interaction at higher levels.

Related to the evaluation of lifelong learning programs conducted through distant education method, there are only a few studies. In the studies conducted up to now, the factors which determine the satisfaction levels weren't diversified into various components. In this study, the satisfaction level of the Spiritual Guidance Education Program among the students which is one of the WBDE application was evaluated in terms of aesthetical and deductive dimensions. Since spiritual guidance program is a factor of distant education due to its content, it is important to evaluate the sufficiency of the aforementioned program.

As is known, they are main evaluation criteria which may influence the satisfaction levels among the learners in the dimensions of teaching designing and technical dimensions in the WBDE application. From this aspect, the evaluation of the Spiritual Guidance Education program in terms of teaching designing and technical dimensions will provide resource to the researchers related to the issue and contribute to the efforts to fill in the gaps in the aforementioned field. Nevertheless, studies containing deep interviews which can reveal the experience and opinions of the students may provide positive contributions to the design and application of WBDE environments.

REFERENCES

Akbulut, K. (2017). Article. *Journal of Financial Solution, 27*, 219-225.

Akca, M., & Aydogan, T. (2015). Online Database Editor Design for Web-Based Distance Learning. *Pamukkale University Journal of Engineering Sciences.*

Alomyan, H. (2017). *A Conceptual Framework for Web-Based Learning Design.* International Association for Development of the Information Society.

Bell, D. (1973). The Coming of Post-Industrial Society: A Venture. In *Social Forecasting*. Harper Colophon Books.

Ben-Chayim, A., & Offir, B. (2019). Model of the Mediating Teacher in Distance Learning Environments: Classes That Combine Asynchronous Distance Learning via Videotaped Lectures. *Journal of Educators Online, 16*(1), n1. doi:10.9743/jeo.2019.16.1.1

Cakmak, A. C. (2013). Evaluation of Distance Learning Service by Students: An Application at Karabuk University. *Istanbul Commerce University Journal of Social Sciences, 12*(23), 263-287.

Castells, M. (2005). *Information Age: Economy, Society and Culture - Rise of Network Society* (E. Kilic, Trans.). İstanbul Bilgi University Publications.

Cheng, F. F., Chiu, C. C., Wu, C. S., & Tsaih, D. C. (2017). The influence of learning style on satisfaction and learning effectiveness in the asynchronous web-based learning system. *Library Hi Tech*, *35*(4), 473–489. doi:10.1108/LHT-12-2016-0151

Cidral, W. A., Oliveira, T., Di Felice, M., & Aparicio, M. (2018). E-learning success determinants: Brazilian empirical study. *Computers & Education*, *122*, 273–290. doi:10.1016/j.compedu.2017.12.001

Crismond, D. P., & Adams, R. S. (2012). The informed design teaching and learning matrix. *Journal of Engineering Education*, *101*(4), 738–797. doi:10.1002/j.2168-9830.2012.tb01127.x

Duffy, P. (2008). Engaging the YouTube Google-eyed generation: Strategies for using Web 2.0 in teaching and learning. *Electronic Journal of E-learning*, *6*(2), 119–130.

Gulnar, B. (2008). Design, Development and Evaluation Stages of Computer and Internet Supported Distant Learning Programs (Sample of Suzep)]. *Selcuk University Journal of the Institute of Social Sciences*, (19), 259–271.

Hsu, L. (2017). Enhancing college students' satisfaction and learning interest when the instructor uses a web-based platform while teaching. *American Journal of Educational Research*, *5*(1), 18–24.

Jackson, L., Jones, S., & Rodriguez, R. C. (2010). Faculty actions that result in student satisfaction in online courses. *Journal of Asynchronous Learning Networks*, *14*(4), 78–96. doi:10.24059/olj.v14i4.129

Kaysi, F., & Aydemir, E. (2017). Evaluation of interaction dimensions in distance learning processes. *Journal of Social Sciences*, *4*(11), 778–790. doi:10.16990/SOBIDER.3442

Korkmaz, O., Cakir, R., & Tan, S. S. (2015). The Effect of Students' Readiness for E-Learning and Satisfaction Levels on Academic Success]. *Journal of Kirsehir Education Faculty*, *16*(3).

Lee, K. (2018). Everyone already has their community beyond the screen: Reconceptualizing online learning and expanding boundaries. *Educational Technology Research and Development*, *66*(5), 1255–1268. doi:10.100711423-018-9613-y

Lee, S. J., & Reeves, T. C. (2007). Edgar Dale: A significant contributor to the field of educational technology. *Educational Technology*, *47*(6), 56.

Metin, A. E., Karaman, A., & Sastim, Y. A. (n.d.). Student Perspectives on Distance Education and the Assessment of the Efficiency of Distance Education English Courses: Banaz Vocational College. *Karabuk University Journal of Social Sciences Institute, 7*(2), 640-652.

Nichols, M. (2016). Reading and studying on the screen: An overview of literature towards good learning design practice. *Journal of Open, Flexible, and Distance Learning, 20*(1), 33–43.

Sahin, T. Y., & Yildirim, S. (1999). Educational Technologies and Material Development. Academic Press.

Virtanen, M. A., Kaariainen, M., Liikanen, E., & Haavisto, E. (2017). The comparison of students' satisfaction between ubiquitous and web-based learning environments. *Education and Information Technologies, 22*(5), 2565–2581. doi:10.100710639-016-9561-2

Yang, X., Guo, X., & Yu, S. (2016). Student-generated content in college teaching: Content quality, behavioural pattern and learning performance. *Journal of Computer Assisted Learning, 32*(1), 1–15. doi:10.1111/jcal.12111

Zhang, D., Zhou, L., Briggs, R. O., & Nunamaker, J. F. Jr. (2006). Instructional video in e-learning: Assessing the impact of interactive video on learning effectiveness. *Information & Management, 43*(1), 15–27. doi:10.1016/j.im.2005.01.004

Chapter 7
New Features of Learners in Education:
Digital Awareness, Digital Competence, and Digital Fluency

Figen Kilic
Mersin University, Turkey

Ismail Karakuş
Mersin University, Turkey

ABSTRACT

Throughout the recent decades there have been rapid developments and changes in terms of science and technology. From now on, individuals want a flexible education format in which time and place are changeable in line with their desires and concerns. Particularly from the perspective of lifelong learning, digitalization can be seen as an important element which forces individuals to change. Some of these are digital awareness, digital competence, and digital fluency. As one of the digital terms, digital competence includes the effective, efficient, critical, flexible, and reflective structuring of information for business, learning, and socializing, while the latter digital fluency is the ability to know when and where to use technology. The main problem that stands out is adapting education programs according to the new developments. The restructuring of the program development activities, so as to raise individuals with changing and developing knowledge, skills, and attitudes, acts like a spotlight which highlights the critical role of program development in the education system.

DOI: 10.4018/978-1-7998-4534-8.ch007

INTRODUCTION

Technological developments in the digital world are known as the concepts that transform human lives. Particularly, just after the spread and development of portable and mobile technologies, it has become virtually impossible for people and institutions to follow developments in many fields including education. For education, it can be said that a digital process, in which developments occur at a rapid speed, has been going on. Developing Information and Communication Technologies (ICT) contributed to both information levels and accessibility of information (Erkan, 1988). It is widely known that validity duration of information has been gradually decreasing, and information needs to be continuously updated. Thanks to digital technologies that are constantly evolving, information has been accepted as the main power and resource in all other fields, especially education, and this has paved the way for a new "digital" society based on information. Therefore, it is very important to make our students adapt to this digital society of the future.

Since individuals with digital competence need to be interactive and productive in the digital environment, they are also expected to be conscious citizens of the digital world. The ability of the individuals to obtain new knowledge and skills in line with adapting to the changes and transformations in technology depends on their flexibility and fluency throughout their lives. The changes and developments in digital technologies radically altered the functioning of society, particularly the way of learning, communicating and working. Thus, the concept of 'Digital' has become an important part of everyday life in the context of skills that can be acquired in the 21st century. Digitalization in education leads to many easy, active, efficient and result-oriented methods in terms of individuals, society and institutions (Parlak, 2017). In such a context, the increasing digitalization of today's world and school life has significantly changed the conditions of education. It can be claimed that it is very significant to restructure and raise all stakeholders of education, students in particular, with digital consciousness.

Thanks to the abovementioned feature, digitalization emerges as one of the most important global trends of the 21st century, and the concept of "digital education" emerges as new technological developments and its reflections on education increase day by day. 'Digital education' refers to the processes of learning and teaching digital technology and the use of digital technology in education (Bayne & Ross, 2011). For this point of view, Prensky (2001) emphasizes that the new generation spends all their lives with computers, tablets, video games, smartphones and all other tools of the digital age. The author defined these digital children of the digital age as "digital natives" and regarded today's students as individuals who speak the digital languages of the internet, video games and computers as if these languages were their mother tongues. Jones et al. also (2010) stated that these digital natives are growing with computers and the Internet, and they have a talent with high skill levels while using new technologies.

On the other hand, it can be said that the new generation is different from all other generations in that the new generation can think differently and learn differently thanks to their constant experiences of technological tools in all areas. As Gracious and Shyla (2012) states, "Digital everything, which enables us to obtain information, to communicate with each other, to influence the environment by using electronic or digital tools". The authors also define digital learning technologies as "Electronic networks including computer learning methods related to technical protocols.". Some researchers in the field also put forward the idea that the growing generation using digital technologies has developed complex technological abilities and new cognitive styles (Dede, 2005; Howe and Strauss, 2000; Oblinger, Oblinger and Lippincott 2005; Prensky, 2001; Tapscott 1998; Veen and Vrakking, 2006 among others). In that sense, the aim of today's digital education is to raise individuals who can use multimedia, who are familiar with technology and who know how to access information (Kılıç, 2014). It is also widely accepted that the use of digital technologies has become an integral part of education, and they directly affect the learning-teaching process (Balamuralithara and Woods, 2010; Çalık & Çınar, 2009; Fidalgo-Neto, Tornaghi, Meirelles, Berçot, Xavier, Castrove Alves, 2009; Göktaş, Yıldırım and Yıldırım, 2008 among others). Hence, it is of high importance to restructure the education system to meet the needs of the new generation that closely follows the technology. Being able to create and sustain this value among individuals also requires a lifelong digital learning process.

Lifelong learning is one of the most important factors which reinforces the idea of digital transformation to individuals. The purpose of lifelong learning is to allow individuals actively participate in all stages of economic and social life so that they can adapt to the information society and control their lives in this society in a better way (Tamer, 2014). The need to adapt to the digital age makes lifelong learning nearly compulsory. On this topic, Coutinho (2007) emphasizes that digital tools play an important role in creating rich learning environments, responding to different learning styles of students, transferring what is learned and supporting higher-order thinking with lifelong learning while offering students the opportunity to compare real-life problems with digital problems.

Via the use of digital tools (computer, mobile tools, internet, Web 2.0, etc.) used in education in the classroom, ways of accessing information among students have also changed, and it has become important for students to access information anytime with the help of their digital competencies (Dodd, 2015). Bright (2017) highlights that individuals of the digital age have free personalities, prejudiced thoughts, unlimited imaginations, broad perspectives, sophisticated aesthetic feelings, rationalist ideas, devotion to science, sports and arts and practical lifestyles; therefore, they are far away from traditionalist structures. The author asserted the idea that students should both use technology and integrate skills with their learning-teaching processes. All these ideas shed light onto the idea that new digital developments directly affect students, teachers and learning environments (Akpınar, 2003). Therefore, it is possible to assert that students, as one of the most important stakeholders of education, have many new types of knowledge, skills and attitudes. It can be said that the features described below are extremely important for the learners, especially in the process of adapting to the digital world, considering that education is digitized.

MAIN FOCUS OF THE CHAPTER[1]

On condition that the individuals are aware of developments and changes in the world they live in, their adaptation to the process can be described as awareness. Besides, being actively involved in the digital society is important to have digital competencies and to solve the problems effectively and creatively. It would be beneficial to gather the features under the titles mentioned below to provide a general perspective.

1. Digital Awareness

Awareness is a frequently used concept in daily life. When we look at its usage in digital environment, it can be observed that it is used for activities aimed at meeting personal needs such as shopping, banking, debt payments, socializations, social shares, online and offline games. (Sürmelioğlu and Seferoğlu, 2019). While these processes are being done, thousands of records can be created digitally. If a user is not aware of this situation, the level of awareness may appear as an important problem in our digitalized life. Hence, awareness of the responsibilities regarding the functioning of our digitalized lives and direction of imdividuals in this respect may come out as another 21st century skill.

Wang (2012) refers to the term "Digital Awareness", and the author emphasizes that it is used in the sense of working with digital tools and environments in many fields such as research, communication, entertainment. Awareness in digital environments plays an important role in ensuring effective digital communication because learners should be aware of the behavior and responsibilities of the people they interact with (Idrus, Abidin, Hashim, and Omar, 2010). Therefore, the studies carried out (Costa, Sousa, Rogado and Henriques, 2017; Krasna, Bratina and Bedrac, 2011) emphasize that digital tools should be used safely, healthily and efficiently. Although some research has been done about the health and safety of learners while using digital tools, little is known about awareness of individuals about these results (Cernikova, Smahel and Wright, 2018). However, exposure to the negative effects of digital media is widespread. In the studies of Cernikova, Smahel and Wright (2018), awareness of children (between the ages of 9-16) and their health-related awareness of the use of digital media were investigated. According to the research, it was displayed that the awareness of the learners about the effects of technology on health can often be affected by excessively distressed / direct content.

In a similar vein, problems such as excessive use of technology, cyberbullying, internet or game addiction, meeting strangers online, headache and abdominal pain, eye problems and poor eating habits, internet addiction, mental fatigue and aggression may frequently occur. For this reason, it can be said that the concept of "digital awareness" does not solely mean that individuals can only follow the developments in digital technology and use digital tools. It is also of high importance to note down that future students act with this awareness and use digital tools correctly and beneficially in the learning-teaching process. When the studies are examined, it can be understood that the learners generally have high proficiency in ethics and usage areas while they have little information about practices (Tekerek and Tekerek, 2013; Yılmaz, 2015; Wang, 2012). Based on these data, the necessity of educating learners about digital awareness is an important issue. Another concept, which is as important as digital awareness, is digital competence.

2. Digital Competence

Digital competence started to be discussed by the European Parliament and Council (European Council, EC) as one of the core competencies for 'lifelong learning' in 2006, and the concept was summarized by European Union (European Union, EU) as " Digital competence includes reliable and critical use of information society technology for business, entertainment, learning and communication." (Sekret and Kommers, 2014; Vieru, 2015). In this sense, digital competence appears as a relatively new concept that has not been well defined, and as a concept of which scope and boundaries have not been determined yet. Although there are different definitions, it is possible to say that the meanings, in which the concept is attributed, are versatile and rich.

Digital competence refers to the skills and literacies that an average citizen needs to learn in the digitized information society (İlomaki et al. 2016). There are many definitions related to digital competence in the literature. As Lennon, Kirsch, Davier, Wagner, Yamamoto (2003) states, it means "Individuals' ability, attitude and ability to use digital technology and communication tools appropriately to access, manage, combine, build and evaluate information, to communicate with others, to engage and effectively participate in society." According to European Commission (EC, 2006), the terms means "the ability of individuals to ensure the use, storage and production of information through the computer, and to communicate and collaborate with relevant stakeholders over the Internet.". As Ferrari (2012) highlights, the term is equivalent to "the set of information, skills, attitudes, talents, strategies and awareness required not only for doing work through ICT and digital media but also for solving problems; managing information; contacting; cooperating; creating / sharing content and information in effective, efficient, appropriate, critical, creative, autonomous, flexible, ethical, reflective ways for the areas such as business, entertainment, participation, learning and socialization ". These skills enable individuals to become digital citizens in a safe way, to interact, collaborate and produce in a digital way and to trust themselves in numerical thinking (Çubukcu, Tosuntaş & İnci, 2018).

Individuals, who are digitally sufficient, can think creatively and critically, improve themselves digitally, and have technological awareness and technical understanding (Ferrari, 2013). As Gallardo-Echenique et al. (2015) stated, the appropriate acquisition of digital competence, highlighted from a holistic and liberating perspective, is an important key to active participation in contemporary society. Individuals with digital competence are expected to be conscious citizens in the digital world as they interact and communicate efficiently and effectively. They are expected to be productive in the digital environment. Ilomäki, Kantosalo and Lakkala (2011) stated that digital competence includes some skills. The authors point out:

- Technical skills to use digital technologies,
- Ability to use digital technologies functionally in various activities and daily life,
- Ability to critically evaluate digital technologies,
- The motivation to participate in digital culture.

Çubukcu, Tosuntaş and İnci (2018) state that digital competence draws a comprehensive road map for the training of students as digital citizens, and it is also stated that it has an integrative role in society (Gallardo-Echenique et al., 2015). It is quite clear how important this role can be in the development of societies. In addition, with the effective use of ICT, being an active member of the society and training of competent individuals, which are two of the most basic functions of education, are once again revealed.

In this context, it is of high significance for learners to involve in digital societies and to produce their own digital materials by applying the information they have learned actively.

One of the important features of being an active member of the society is the concept of digital fluency.

3. Digital Fluency

The studies in the literature have proposed different definitions of digital fluency as a skill characterized by 'lifelong learning, creativity, critical thinking, self-management', and as a skill that all individuals should acquire in the 21st century (Sparrow, 2018; Pinho & Lima, 2013, Miller & Bartlett, 2012). Digital fluency, as proposed by Hsi (2007), is defined as "Learners acquire and demonstrate actions to obtain, design, evaluate, criticize, synthesize and develop digital media content, messages and other electronic expressions using digital tools". According to Miller and Bartlett (2012), it is "the combination of critical thinking and information literacy skills". A study in a similar vein, Pinho and Lima (2013) point out that it is "the ability to adapt to the transformations in technology in order to obtain new information that may be encountered throughout life". Niessen (2013) also includes "a talent for when and why we use the digital media we choose."

As a recent definition, Liu, Li, & Ye (2018) have proposed that it is composed of "critical thinking, information comprehension skills and the use of private information". According to Sparrow (2018), digital fluency is the ability to create something new with technological tools, and it can be considered as a collection of fluencies that has unlimited components such as 'curiosity, communication, creativity, data management and innovation' which are still developing. Sparrow (2018) proposes the components of digital fluency as follows. (see Figure 1).

Figure 1. Components of the digital fluency

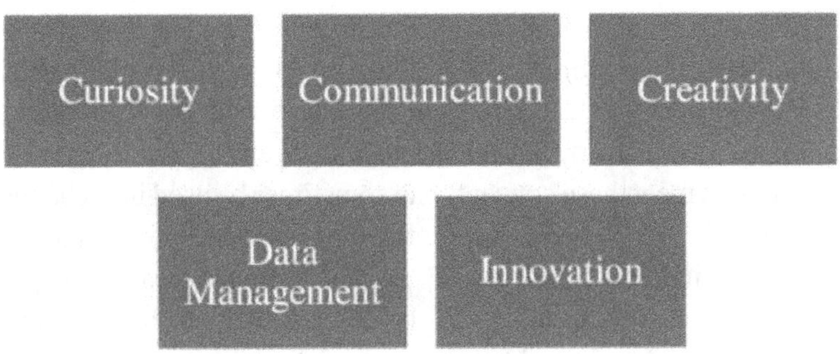

- Curiosity includes asking questions and the desire to answer these questions.
- Communication fluency is the ability to transfer new information among different communities and choose an environment that is most appropriate and effective for a particular audience.

- Creativity is a deep understanding of how information is created and how information is used to do something new.
- Data management is the capacity to use data sets, while pushing the limits of what technology can do during data processing, to make new and effective decisions..
- Innovation involves realizing that failure is a valuable part of the learning process. In order to innovate, students require taking risks, failing, learning from these failures and developing new ideas.

Sparrow (2018) defines digital fluency as "the ability to use technology for creating new problems, thinking critically and solving complex problems via social intelligence." The author explained that digital fluency requires new media literacy with excellent communication skills to address the problems and concerns we may face today and in the future. White (2013) also conducted a study that emphasizes how important it is for the students to acquire these skills for adapting to life. The National Research Council (NRC, 1999) stated that digitally fluent students can use information technologies in accordance with their personal and professional activities to learn, distinguish, evaluate, adapt to changing technology. Thus, they can use digital tools effectively throughout their lives. In that sense, they are individuals who take part in lifelong learning experiences. Demir (2018) also highlighted the importance of lifelong learning and digital fluency by stating that creating and sustaining digital fluency among individuals is a lifelong process. In this respect, it is vital for the learners, who have very important missions in building the future, to have digital fluency within the framework of lifelong learning.

Taking the explanations in the literature into account (Spencer, 2015; Aesaert, van Nijlen, Vanderlinde and Van Braak, 2014; Briggs & Makice, 2012; Ferrari, 2012), it can be said that there are no clear definitions that can reveal the differences between the concepts of 'digital awareness, digital competence and digital fluency', It is seen that the features that meet these concepts can sometimes be used interchangeably. In this context; digital competence can be defined as 'knowledge, skills and attitudes required to use digital tools effectively.' Digital awareness can be explained as 'being aware of the events occurring in the environment of the individual about digital technologies and tools and being aware of digitalization with the ability to perceive.', while digital fluency is the ability of the individual to adapt to digital technologies and experience learning within the context of lifelong learning." Due to the abovementioned discussions on the definitions of the concepts from the literature, it can be said that the definitions and explanations for these three concepts are intertwined (Sparrow, 2018; MEB; Spencer, 2015; Niessen, 2013)

In a study in a similar vein, Briggs and Makice (2012) describe the characteristics of digitally fluent individuals: "Being aware of when and how to use digital tools

(1), using digital tools very effectively and making digital tool use an automated skill (2), developing new technologies as digital technologies change and develop and being adaptable (3) ". According to this explanation, one of the characteristics of digitally fluent individuals might be related to digital awareness, the second to digital competence and the third to digital fluency.

Figure 2. The Relationships among Digital Fluency, Digital Competence and Digital Awareness

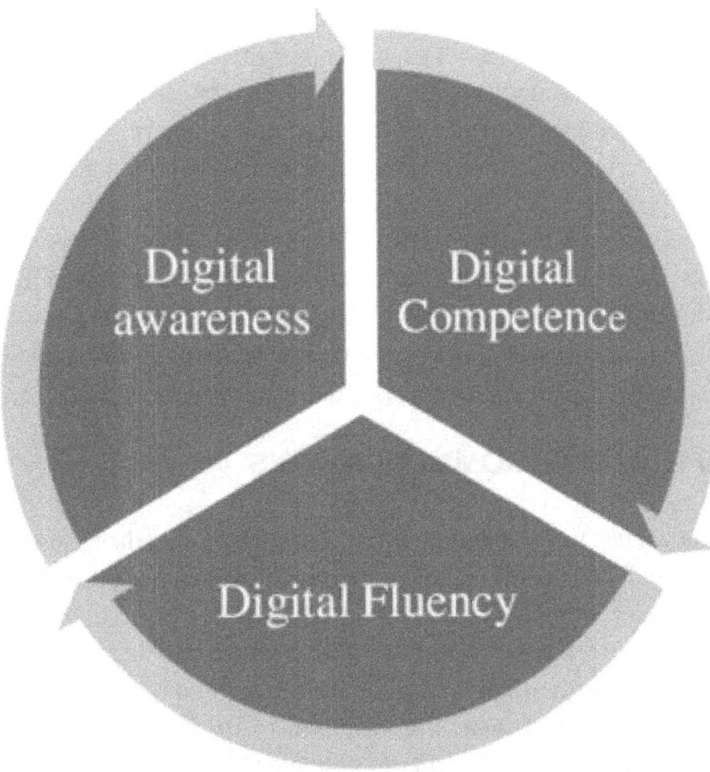

In the USA, the standards set by the International Society for Technology in Education (ISTE) emphasize that students should use digital tools and technologies effectively in the fields of digital citizenship, communication, collaboration, innovative design and information creation (Çebi and Reisoğlu, 2019). It is claimed that one of the eight competencies in Europe is digital competence (EC, 2007). In the light of this idea, DigComp framework has been started to be developed within the scope of the Digital Competency Project in order to comprehend and develop digital competence in Europe in a better way (Vuorikari, Punie, Carretero Gomez & van den Brande,

2016). Unlike other frameworks, DigComp includes higher-order thinking skills by taking 21st century skills into account (Janssen, Stoyanov, Ferrari, Punie, Pannekeet & Sloep, 2013). Within the framework of the abovementioned Digital Competency Project, the aim is to make individuals obtain 21st century skills. Therefore, it can be said that the 21st century skills in the project are related to digital fluency and digital awareness skills.

Based on all the discussions, it is possible to claim that individuals, who are exposed to traditional methods, philosophies and approaches of education (particularly when they are raised for our digital world) may be inadequate. Since there is a new process, in which academic and institutional education is digitized and redesigned in every respect, all stakeholders of education, especially teachers and students, are affected by this change generally. In such a context, both teachers and students have to adapt to this digital transformation and develop themselves regardless of their age, which may come out as an important problem. One of the most basic ways to deal with this problem is to start the change with the teachers who will build the future. Teachers require a spesific level of interaction and competence in digital environments, awareness about digital tools and applications and an ability to think flexibly within the framework of lifelong learning. Furthermore, they need to obtain the ability of learning to learn, be fluent and train students in this framework.

SOLUTIONS AND RECOMMENDATIONS

It is a well known fact that the digital developments are felt in the field of education as in all fields, and they are closely related to the all education stakeholders. Hence, the most important elements of education stakeholders emerge as students and teachers. Furthermore, it is also widely accepted that the new generation needs to be equipped with today's information, skills and attitudes, and it is of great importance for both the education system and the social sense. To serve for this aim, some of the 21st century skills for the students to obtain are expressing themselves, using a foreign language effectively, having research skills, adapting to developing and changing technologies and using them effectively. However, most of the researches show that not all students around the world have these basic skills, or they have the skills with an ineffective usage. In this case, it is extremely important to find ways to raise individuals with these skills. The reason is that there is a new process in the academic and institutional senses in which societies are digitalized and redesigned in all aspects of education. From the general perspective, all the stakeholders of education, especially teachers and students, are affected by this change / transformation. Therefore, it is considered as an important problem for both teachers and students to cope with this digital transformation regardless of their age.

In order to solve this problem, reports / studies on the effects of technology on children should be examined. Particularly, parents and teachers should be encouraged to have these critical perspectives by adopting a more balanced perspective on the impact of technology on health. Beyond these, organizing curriculum for digital awareness at schools might be one of the most important solutions. In the study the author conducted in Scandinavian countries, Siddiq (2018) states that education programs for digital awareness do not have a sufficient structure inasmuch as their competence areas do not support a conceptual hierarchy. It was also noted that the lack of explanations regarding teaching aspects (eg assessment, teaching procedures and materials) could affect the integration of the program. In the study of Tekerek and Tekerek conducted in Turkey (2013), the awareness level of elementary and high school students on ethical issues was found out to be sufficient, while the same awareness level was found to be low in rules and subjects requiring information. The reason for this lies behind the fact that information and computer security awareness training and activities were insufficient, and it was suggested that the number of courses on this subject should be increased. In another study conducted on the same topic, Wang (2012) emphasized that the universities should actively support interdisciplinary discussions, including both face-to-face seminars and opportunities and sharing ideas in online forums, among established and emerging researchers (doctoral students) so as to raise awareness. These discussions should not be limited to practical tips and strategies for using digital technology. They should also focus on how digital tools change information generation and sharing practices. The last but not the least, it is a fact that many teachers have inadequate information and digital fluency levels about how digital technologies can be effectively integrated into the classroom environment in the learning-teaching process (Chigona, 2018). On this topic, there are studies showing that digital fluency is a motivational factor in foreign language teaching as a competence needed in a remarkable and active learning process. Therefore, it is vital to give institutional support to students and teachers to make them digitally fluent (Pinho & Lima, 2013).

Based on the concepts and discussions on the literature, it can be summarized that digital fluency should not only be considered as a technical concept; instead, it should also focus on its social and psychological aspects. The possibility of experiencing digital tools will possibly increase as the levels of income and education level increase. Due to this fact, it is necessary to combine daily life and education practices in addition to creating programs for critical thinking. With such combinations, individuals, who are open to innovations, manage complexity, produce solutions, adapt quickly to technological transformation, are going to take part in our daily lives in the future. The most important proposal that can be made based on all these inferences is that there is a necessity to organize digitalization as a training program at schools. In a study conducted by White (2013), the author stated that "digital fluency" should be taught as a course, which emphasizes the importance of acquisition of these skills by students and their adaptation to life. These subjects are listed below.

- History of Internet
- Socialization
- Critical Thinking
- Design Skills
- Digital Fluency
- Ethics
- Technology
- Identity and Privacy
- Security
- Digital Rights and Copyright
- Collaboration, Communication, Problem Solving and Research Strategies
- Project management
- Acceptable Behavior

Taking the contents of the courses mentioned above, the necessity of raising responsible individuals, who are creative, aware of what they do, active in the society, is revealed once again.

IMPLICATIONS FOR FUTURE RESEARCH

In the light of the information discussed in the chapter, it can be asserted that many new concepts have emerged in the process of adaptation to the developing and changing world. It is also known that lifelong learning has gained importance in the 21st century, and people should have many skills. Wagner (2008) puts forward the idea of acquiring 21st century skills. The author particularly highlgihts critical thinking, problem solving, accessing and analyzing information, systems and interpersonal cooperation, leadership, entrepreneurship, agility, adaptability, effective oral and written communication, curiosity and imagination. Indeed, the moment all of these skills are considered, individuals are expected to collaboratively produce something new through effective communication, use modern ways effectively while doing all these and contribute to the development of both themselves and the society. Definitely, the ways to obtain these may be welcoming new concepts about "digitalization", which has started to enter our daily life rapidly, and developing an effective training program that will enable these concepts to be used as operational concepts. In this era, it is understood that both relational and descriptive explanations are needed for the concepts of "digital awareness, digital competence and digital fluency". Turkey, as a country in which the identification of common and different aspects of these concepts has begun to be understood, has to determine how these concepts play an important role for the curricula. In line with the data obtained from all over the world, undoubtedly, the development of training programs in this context should be among the most important issues that program development specialists attach importance. Developing and testing an effective and qualified program proposal in order to obtain the skills expected from learners and the implementation of such a program at schools are crucial for the future.

CONCLUSION

The 'Coronavirus / Covid-19' pandemic, has not only detoriated the life conditions in all countries but also it has made many activities in everyday life stop. Naturally, education activities have been stopped in many countries, and educational activities have been continued using digital tools in remote / online digital media. This situation caused us to leave aside the traditional methods and tools we are accustomed to and take part in digital platforms, which has also led to the necessity to have a lot of skills. From the usage purposes of the platforms, it has become necessary to be digitally aware in many areas such as security, efficiency, ethics and health. Studies on this issue show that individuals do not attach enough importance to information security, personal awareness of information security awareness, and meanwhile, students are not provided with educational opportunities in which they can increase their knowledge and awareness (Dönmez, 2019). Atwell also emphasized that individuals in the digital age are both ignorant in the digital sense. Individuals, who lack digital skills, are excluded, and new ways to prevent this exclusion must be found (As Cited in;Siebert, 2010).

Lifelong learning is not limited to schools. Individuals can access to education at home, at work, in all areas of life. It is one of the basic concepts that shows that learning can be continued without any obstacle regardless of age, social / economic status and education level (MEB, 2020). Considering that the concept of lifelong learning is a basic skill in order to keep up with the changes in the rapidly changing and developing socio-cultural life of the era, it is also very important to examine the skills of the students in this field and acquire these skills.

Based on all of these, it is of high significance to raise awareness of the "digital" concepts encountered in the 21st century, to develop our awareness individually and to develop educational programs. Thus, individuals become active members of the society by having the skills included in these concepts.

REFERENCES

Aesaert, K., van Nijlen, D., Vanderlinde, R., & van Braak, J. (2014). Direct measures of digital information processing and communication skills in primary education: Using item response theory for the development and validation of an ICT competence scale. *Computers & Education*, *76*, 168–181. doi:10.1016/j.compedu.2014.03.013

Akpınar, Y. (2003). Öğretmenlerin yeni bilgi teknolojileri kullanımında Yükseköğretimin etkisi: İstanbul okulları örneği. *The Turkish Online Journal of Educational Technology*, *2*(2), 79–96.

Balamuralithara, B., & Woods, P. C. (2010). An Investigation on Adoption of the Engineering Simulation Lab Exercise: A Case Study in Multimedia University, *Malaysia. Computer Applications in Engineering Education*, *1*(22), 34–46.

Bayne, S., & Ross, J. (2011). *Digital native'and 'digital immigrant'discourses: a critique. In digital difference*. Brill Sense. doi:10.1007/978-94-6091-580-2_12

Briggs & Makice. (2012). *Digital Fluency: Building Success in the Digital Age*. Socia Lens.

Çalık, D., & Çınar, Ö. P. (2009). Geçmişten günümüze bilgi yaklaşımları bilgi toplumu ve internet. In XIV. Türkiye'de İnternet Konferansı (pp.77-88). Bilgi Üniversitesi.

Çebi, A., & Ve Reisoğlu, İ. (2019). Öğretmen adaylarının dijital yeterliklerinin geliştirilmesine yönelik bir eğitim etkinliği: Böte ve diğer branşlardaki öğretmen adaylarının görüşleri. *Eğitim Teknolojisi Kuram ve Uygulama*, *9*(2), 539–565.

Cernikova, M., Smahel, D., & Ve Wright, M. F. (2015). The impact of digital media on health: Children's perspectives. *International Journal of Public Health*, (60), 131–137. PMID:25601331

Chigona, A. (2018). Digital fluency: Necessary competence for teaching and learning in connected classrooms. *The African. Journal of Information Systems*, *10*(4).

Costa, C., Sousa, C., Rogado, J., & Henriques, S. (2017). Playing digital security: Youth voices on their digital rights. *International Journal of Game-Based Learning*, *7*(3), 11–25. doi:10.4018/IJGBL.2017070102

Coutinho, C. (2007). Infusing technology in pre-service teacher education programs in portugal: A study with weblogs. In *Proceedings of the 18th International Conference of the Society for Information Technology & Teacher Education*. Chesapeake, VA: AACE.

Çubukçu, Z., Tosuntaş, Ş. B., & İnci, T. (2018). Eğitim programları aracılığıyla dijital yeterliklerin kazandırılması Dijital Yeterlikler Çerçevesi örneği. Presented at the *International Symposium on Multidisciplinary Academic Studies*, Berlin.

Dede, C. (2005). Planning for neomillennial learning styles. *EDUCAUSE Quarterly*, *28*(1), 7–12.

Demir, K. (2018). Öğretmen adaylarının dijital akıcılıklarının incelenmesi. Doktora tezi, Anadolu Üniversitesi Eğitim Bilimleri Enstitüsü, Eskişehir.

Dodd, W. D. (2015). *Teaching and Learning in the Digital Age*. https://webcpm.com/articles/2015/09/01/teaching-learning-technologies.aspx

Dönmez, G. (2019). *Lise Öğrencilerinin Bilgi Güvenliği Farkındalığı İle Dijital Okuryazarlığı Arasındaki İlişkinin İncelenmesi*. Yüksek Lisans Tezi. Hacettepe Üniversitesi Eğitim Bilimleri Enstitüsü.

Erkan, H. (1998). *Bilgi Toplumu ve Ekonomik Gelişme*. Türkiye İş Bankası Kültür Yayınları.

European Commission-EC. (2006). The Key Competences for Lifelong Learning – A European Framework. Official Journal of the European Union, 394(1).

European Commission-EC. (2007). Key competences for lifelong learning - A European framework. Luxembourg: Author.

Ferrari, A. (2012). Digital competence in practice: An analysis of frameworks. Sevilla, Spain: Academic Press.

Ferrari, A. (2013). *DIGICOMP: A framework for developing and understanding digital competence in Europe*. Luxembourg: JRC Scientiðc and Policy Reports EUR26036EN.

Fidalgo-Neto, AA, Tornaghi, AJC, Meirelles, RMS, Berçot, FF, & Xavier, LL, Castro, MFA ve Alves, LA. (2009). Brezilya ilk ve ortaokullarında bilgisayar kullanımı. *Bilgisayar ve Eğitim*, *53*(3), 677–685.

Gallardo-Echenique, E. E., de Oliveira, J. M., Marqués-Molias, L., Esteve-Mon, F., Wang, Y., & Baker, R. (2015). Digital competence in the knowledge society. *Journal of Online Learning and Teaching*, *11*(1).

Göktas, Y., Yildirim, Z., & Yildirim, S. (2008). A Review of ICT Related Courses in Pre-service Teacher Education Programs. *Asia Pacific Education Review*, *9*(2), 168–179. doi:10.1007/BF03026497

Gracıous, A., & Shyla, A. J. (2012). Multıple ıntellıgence and dıgıtal learnıng awareness of prospectıve teachers. *Turkish Online Journal of Distance Education-TOJDE, 11*(3), 112–118.

Howe, N., & Strauss, W. (2000). *Millennials rising: The next great generation.* Vintage Original.

Hsi, S. (2007). Conceptualizing learning from the everyday activities of digital kids. *International Journal of Science Education, 29*(12), 1509–1529. doi:10.1080/09500690701494076

Idrus, Z., Abidin, Z. Z. S., & Hashım, R. & Omar, N. (2010). Awareness in networked collaborative environment: a comparative study on the usage of digital elements. *Proceedings of the 9th WSEAS International Conference on applıcatıons of computer engineering*, 236-241.

Ilomäki, L., Paavola, S., Lakkala, M., & Kantosalo, A. (2016). Digital competence– an emergent boundary concept for policy and educational research. *Education and Information Technologies, 21*(3), 655–679. doi:10.100710639-014-9346-4

Jackson, S. (2006). Learning to live: The relationship between lifelong learning and lifelong illness. *International Journal of Lifelong Education, 25*(1), 51–73. doi:10.1080/02601370500309501

Janssen, J., Stoyanov, S., Ferrari, A., Punie, Y., Pannekeet, K., & Sloep, P. (2013). Experts' views on digital competence: Commonalities and differences. *Computers & Education, 68*, 473–481. doi:10.1016/j.compedu.2013.06.008

Jones, C., Ramanau, R., Cross, S., & Healing, G. (2010). Net generation or digital natives: Is there a distinct new generation entering university? *Computers & Education, 54*(3), 722–732. doi:10.1016/j.compedu.2009.09.022

Kilic, F. (2014). Awareness and Cognitive Load Levels of Teacher Candidates towards Student Products Made by Digital Storytelling. *Turkish Online Journal of Distance Education, 15*(3), 94–107. doi:10.17718/tojde.90227

Krašna, M., Bratina, T., & Bedrac, B. (2011). Digital competences in education-digital security. In *INTED2011 Proceedings* (pp. 1634-1641). IATED.

Lennon, M., Kirsch, I., Von Davier, M., Wagner, M., & Yamamoto, K. (2003). *Feasibility Study for the PISA ICT Literacy Assessment. Report to Network A* Educational Testing Service.

Liu, A., Li, Y., & Ye, H. (2018). Digital fluency and social media use: An empirical ınvestigation of WeChat use. *Proceedings of the 51st Hawaii International Conference on System Sciences*, 503-512. 10.24251/HICSS.2018.065

Miller, C., & Bartlett, J. (2012). 'Digital fluency': Towards young people's critical use of the internet. *Journal of Information Literacy*, *6*(2), 35–55. doi:10.11645/6.2.1714

Milli Eğitim Bakanlığı (MEB). (2018). *Öğretim Programları*. http://mufredat.meb.gov.tr/Programlar.aspx

Milli Eğitim Bakanlığı (MEB). (2020). http://hayatboyu.meb.gov.tr/hayat-boyu-ogrenme/

National Research Council, Committee on Information Technology Literacy. (1999). *Being fluent with information technology*. National Academies Press.

Niessen, S. (2013). *What is digital fluency*. http://www.shuananiessen.ca/what-is-digital-fluency/

Oblinger, D., Oblinger, J. L., & Lippincott, J. K. (2005). Educating the net generation. EDUCAUSE.

Odabaşı, H. F. (2000). Akademik eskimişlik ve sürekli mesleki gelişim. *Anadolu Üniversitesi Eğitim Fakültesi Dergisi*, *10*(2), 161–166.

Parlak, B. (2017). Dijital çağda eğitim: Olanaklar ve uygulamalar üzerine bir analiz. *Süleyman Demirel Üniversitesi İktisadi ve İdari Bilimler Fakültesi Dergisi*, (22), 1741–1759.

Pinho, I. D. C., & Lima, M. S. (2013). Teacher's digital fluency: A new competence for foreigh language teaching. *Revista Brasileira de Lingüística Aplicada*, *13*(3), 711–739. doi:10.1590/S1984-63982013005000014

Prensky, M. (2001). Digital Natives, Digital Immigrants. From On the Horizon, 9(5).

Sekret, I. V., & Kommers, P. A. (2014). Conceptual issues of the digital competence development in the framework of the council of the European Union. *Conference proceedings of 10th International Scientific Conference on Distance Learning in Applied Informatics DIVAI2014*, 183-195.

Siddiq, F. (2018, October). *A Comparison Between Digital Competence in Two Nordic Countries' National Curricula and an International Framework: Inspecting their Readiness for 21st Century Education*. Academic Press.

Siebert, D. (2010). *Eğitimde Dijital Devrim*. Deuthsche Welle.

Sparrow, J. (2018). *Digital fluency: Preparing students to create big, bold problems*. https://er.educause.edu/articles/2018/3/digital-fluency-preparing-students-to-create-big-bold-problems

Spencer, K. (2015). *What is Digital Fluency*? [blog post]. http://blog.core-ed.org/blog/2015/10/what-is-digital-fluency.html

Sürmelioğlu, Y., & Seferoğlu, S. S. (2019). Yükseköğretim Öğrencilerinin Dijital Ayakizi Farkındalık ve Yaşantılarının İncelenmesi. *World Journal on educational technology. Current Issues, 11*(1), 48–64.

Tamer, M. G. (2014). Yaşam boyu öğrenme için ulusal yeterlilikler çerçevesi: Avrupa ve Türkiye örneği. *International Journal of Curriculum and Instructional Studies, 3*(5).

Tapscott, D. (1998). *Growing up digital: The rise of the net generation*. McGraw Hill.

Tekerek, M. ve Tekerek, A. (2013). A research on students' information security awareness. *Turkish Journal of Education, 2*(3), 61–70.

Veen, W., & Vrakking, B. (2006). *Homo zappiens, growing up in a digital age*. Network Continuum Education.

Vieru, D. (2015). Towards a multi-dimensional model of digital competence in small-and medium-sized enterprises. In *Encyclopedia of Information Science and Technology* (3rd ed., pp. 6715–6725). IGI Global. doi:10.4018/978-1-4666-5888-2.ch660

Vikipedi. (2020). https://tr.wikipedia.org/wiki/2019-20_koronavir%C3%BCs_pandemisi

Vuorikari, R., Punie, Y., Carretero Gomez, S., & van den Brande, L. (2016). DigComp 2.0: The Digital Competence Framework for Citizens (Update Phase 1: The Conceptual Reference Model). Luxembourg: Publication Office of the European Union.

Wagner, T. (2008). *The global achievement gap: Why even our best schools don't teach the new survival skills our children need-and what we can do about it*. Basic Books.

Wang, K. (2012). *Digital awareness as an aspect of digital literacy*. http://as.exeter.ac.uk/cascade/digital-curriculum/case-studies/digitalawarenessasanaspectofdigitalliteracy/

White, G. K. (2013). *Digital fluency: Skills necessary for learning in the digital age*. Australian Council for Educational Research.

Yılmaz, E. (2015). *Öğretmenlerin dijital veri güvenliği farkındalığı* [Yayınlanmamış Doktora Tezi]. Anadolu Üniversitesi, Eğitim Bilimleri Enstitüsü.

ADDITIONAL READING

Angeli, C., & Valanides, N. (2009). Epistemological and methodological issues for the conceptualization, development, and assessment of ICT-TPCK: Advances in technological pedagogical content knowledge (TPCK). *Computers & Education, 52*(1), 154–168. doi:10.1016/j.compedu.2008.07.006

Fraillon, J., Ainley, J., Schulz, W., Friedman, T., & Gebhardt, E. (2014). Preparing for Life in a Digital Age. The IEA International Computer and Information Literacy Study International Report. Amsterdam: IEA. doi:10.1007/978-3-319-14222-7

Gudmundsdottir, G. B., Hatlevik, O. E., Ottestad, G., & Wiberg, L. K. (2014). *New teachers' digital competence and experiences of ICT in teacher education programmes in Norway. 17th UNESCO-APEID International Conference*, Bangkok, Thailand.

Jenkins, H. (2006). *Convergence culture: Where old and new media collide*. New York University Press.

Meyers, E., Erickson, I., & Small, R. (2013). Digital Literacy and Informal Learning Environments: An Introduction. *Learning, Media and Technology, 38*(4), 355–367. doi:10.1080/17439884.2013.783597

Paniagua, A. (2018). *Why pedagogy matters for innovative teaching*. Education & Skills Today.

Verma, A., & Jha, M. & Mitra, M. (2016). Digital awareness and ınternet usage by school students. *Journal of Psychosomatic Research, 11*(2), 259–270.

KEY TERMS AND DEFINITIONS

Digital Awareness: It is the awareness of the knowledge, skills and attitudes required for individuals to use digital tools effectively.

Digital Competence: It is the ability of individuals to use digital technology and communication tools appropriately to access, manage, combine. and evaluate information. It is also considered as attitude, ability, and ability to communicate with others to build new information and to participate society effectively (Lennon et al., 2003).

Digital Fluency: It is the ability to adapt to the changes in technology in order to obtain new information that can be encountered throughout life (Pinho & Lima, 2013).

Lifelong Learning: Lifelong learning consists of learning experiences that are programmed at home, at work, in the environment we live in, or in educational institutions. People might take part in them by chance, formally or informally (Jackson, 2006).

Chapter 8
Content-Generating Learners' Access to Accurate and Up-to-Date Information

Hakan Kilinc
https://orcid.org/0000-0002-4301-1370
Anadolu University, Turkey

T. Volkan Yuzer
Anadolu University, Turkey

ABSTRACT

This study aims to reveal the characters and responsibilities of content generating learners, as new generation learners, in the learning processes for producing the contents which contain accurate and up to date information. In this study, which was designed as a case study for this purpose, the opinions of the field experts who voluntarily participated in the study were obtained about what features the content generating learners should have at the point of accessing accurate and up-to-date information. The answers given by the field experts to the research questions directed to them were analyzed through the NVIVO 11 package program, and the themes reached were presented within the scope of the study. It is thought that this study will contribute to the related literature in order to ensure that the content generating learners can produce more accurate and more up-to-date content.

DOI: 10.4018/978-1-7998-4534-8.ch008

INTRODUCTION

The change of learning environments and having a technological infrastructure can be seen as a development that changes the responsibilities and qualification limits of learners and instructor participating in these environments (Lambert, Philip ve Nakamura, 2017). The inclusion of technology in learning processes can be seen as an element that will increase the efficiency of learning environments. This view was expressed by Lee et al. (2008); "The use of technology in learning environments motivates learners and causes them to enter learning processes more actively. In this case, it can be said that it is possible to reach the targeted goals in learning environments. Contemporary learning environments are a learning environment that is intertwined with technology, where learners are actively involved in learning processes and can also share their own experiences. " is in line with the expression. Another view expressed in this direction was the expression "Technological inventions have enabled learners to be more active and creative in their learning environments", as stated by Dale and Povey (2009).

Thanks to the developing information and communication technologies (ICT), the lifelong learning need needed in learning environments can be provided (Fischer & Konomi, 2005). It is seen that, with the technological developments, the communication and interaction of individuals with each other has increased. In this way, it can be said that it is easier to access information and to generate information (Kılınç, 2016). By using digital learning environments supported by technology and providing lifelong learning, it has become possible for learners to generate content. At this point, the expression "Using technology is the most important factor that enables learners to generate content" stated by Narayan (2011) shows that technology is an important factor for learners to generate content. From this point of view, it can be said that, thanks to the modern learning environments with technological infrastructure, the concept of content generating learner contributes to the formation of the learner with the learning content. Content generating learners are also defined as learners who create learning content to meet their learning needs individually or by a group of collaborators (Luckin et al., 2005; Luckin, 2006; Sener, 2007).

Digital learning environments can be defined by individuals who communicate and interact with each other or individually as environments that enable the generation of content to achieve learning objectives, while its main feature is shown as enabling individuals to generate content themselves (Luckin et al., 2007). Sener (2007) stated in his study that digital learning environments are a suitable tool that can be used in learning processes. As a reason for this view, "Many online learners aim to be able to produce and consume their generation in non-monetary environments," which Alvin Toffler mentioned in his book "Third Wave" showed his opinion. In this context, digital learning environments can be seen as an element that will improve the academic success of learners. These environments, which enable learners to generate content, can be considered as an important point in the sense that what is learned is permanent.

Learners, who are always available online in the internet environment, become producers rather than passive consumers and help create a more active learning process (Belanche, Casaló, Orús ve Pérez-Rueda, 2020; Klamma et al., 2007). Another conclusion made in the study under The Pew Internet and American Life Project (Lenhart and Madden, 2005) is that approximately 50% (12 million) of Americans in America do not only participate in online learning activities, but also blog their own learning content, It has been created through personal Web addresses and other means. In this context, it can be said that learners can turn their thoughts into products with easy-to-use Web tools. As a result of the content generating learners being constantly active in the internet environment, it is seen that the relationship of the learners with information changes (Lambert ve Zhang, 2019). Content generating learners became active participants in learning activities and took part in the creation of learning content thanks to Web 2.0 (O'Reilly, 2005). Content generating learners have become in a position to create individual learning environments as active content creators who design learning materials with the support of technology.

Another definition of content generating learners that must adapt to the requirements of digital environments has been in the QMPP project (2009) as "active creators of learning materials, personal learning environments and their own learning processes". Another statement stated in line with this idea is that "With Millennium, learners have started to be defined as" consumers "who can generate content and use information in this process," (Lee and McLoughlin, 2007). At this point, it is possible to say that the content generating learners are a producer in terms of their contribution to the learning content and a consumer in the point of using the information they acquire in the process of producing the content. The fact that the content generating learners are in such a position has also brought about a new concept. This concept is expressed as Consumer + Producer = Prosumer (Lee and McLoughlin, 2007; Sener, 2007). Prosumer can also be defined as new generation learners. It is seen as an important factor that the new generation learners, who emerged with the active role of the learners in the learning processes, have access to the correct and updated information. In this context, there is a need for studies to be conducted on how content generating learners should access accurate and up-to-date information while producing their own content. Accordingly, it can be stated that learners who need to reach correct and up-to-date information should act in accordance with 21st century skills and connection theories.

Purpose of the Study

The purpose of this study is to reveal what features the content generating learners should have in order to access correct and up-to-date information. The research questions determined for this purpose are as follows:

- How can content generating learners access accurate information in the content generation process?
- What responsibilities should the content generating learners have in order to understand that they have reached the right information?
- What kind of a path should the content generating learners follow in order to reach up-to-date information?
- In which responsibilities should content generating learners have access to up-to-date information?

METHOD

In this study, the focus of 21st century skills theory and connectivism theory has been handled in a holistic approach in determining and evaluating the characteristics of content generating learners . The entire research process took place as a discovery process, and it was not possible for the researcher to control the facts and events.

Research Model

This study, designed to evaluate the characteristics that content generating learners should have, was designed with a case study, one of the qualitative research approaches. Qualitative research is defined by Yıldırım and Şimsek (2008) as a research in which qualitative data collection techniques such as observation, interview and document analysis are used, and a process is carried out to reveal perceptions and events in a natural and realistic manner. The facts and events that are researched on are considered in their own context and interpreted in terms of the meaning that people attach to them (Altunışık et al., 2010). "The holistic understanding of any phenomenon can only be achieved through multiple perspectives." (Yıldırım and Şimşek, 2008). In this context, in this study, where opinions of different field experts are consulted, it is aimed to obtain a holistic understanding of the concept of content generating learners . The steps of the qualitative research process are shown in Figure 1.

Figure 1. The steps of the qualitative research process (Yin, 2009)

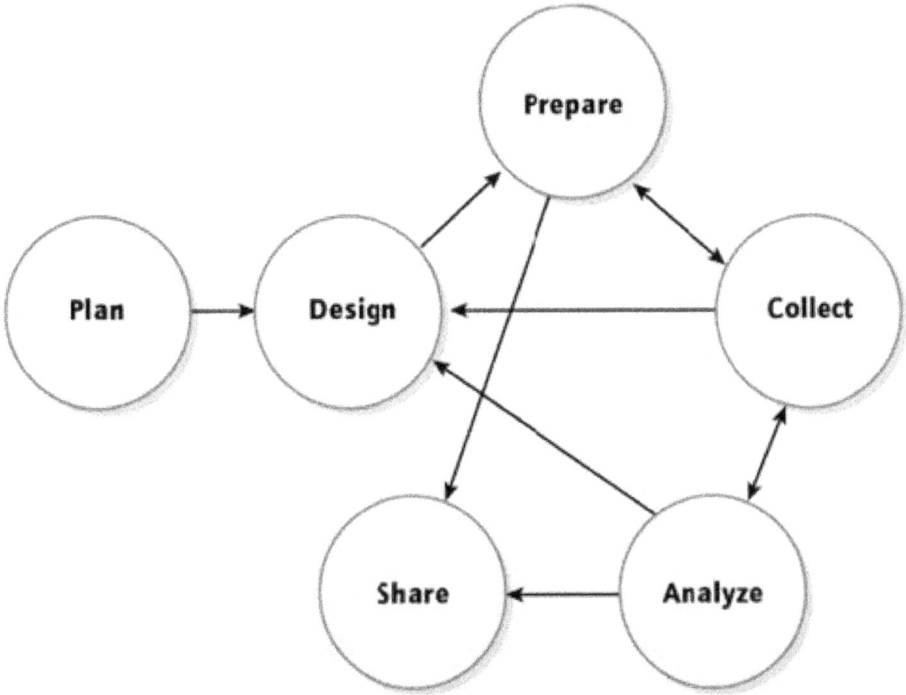

The case study that forms the pattern of this research is based on in-depth analysis of any event, individual or process based on data (Creswell, 2007). According to Yin (2003), case study is a method that investigates the phenomenon in the existing natural environment, is used in cases where the existing situation and the environment in which it is located are not separated by precise lines and multiple data sources are available. Case studies are defined as a research method that works in a real case within its real life framework, where the boundaries between the case and the environment in which it is located are not evident and there is more than one source of evidence or data (Yin, 1984).

It is stated that case study includes (1) type of research question, (2) control of the research on real behavioral events, and (3) advantages and disadvantages depending on contemporary phenomena in relation to historical phenomena, in the process of in-depth analysis and interpretation in which the answers to the questions of how and why are sought. It is stated that it is a preferred method in cases where it has little control over events (Yin, 2002).

In this study, which was designed with a case study, which is one of the qualitative research approaches, the interview technique, one of the qualitative research methods, was used. Interview technique is one of the qualitative research methods used to analyze one or more specific situations by analyzing them in depth (Creswell, 2013). Interview is considered as a powerful data collection method because of the interaction between the researcher and the data source and facilitating the clarification, explanation, detailing of the data collected for the researcher (Yıldırım & Şimşek, 2005). In this research conducted with experts who have online teaching experience in higher education, semi-structured interview technique, one of the interview techniques, was used. Semi-structured interview provides more flexibility compared to structured negotiations (Türnüklü, 2000; Fırat et al., 2014). The opinions of three other experts were taken and triangulation was performed in order to conduct the credibility study of the data collected through the semi-structured interview form used in this study.

Participants

In the research conducted, purposeful sampling was used in determining the participants. Purposeful sampling is one of the useful sampling methods in qualitative research (Creswell, 2005). In the purposeful sampling, the prerequisite for the field experts to be interviewed was to have a PhD in distance education and to have taught in digital learning environments such as distance learning environments for at least one year. The search for the condition that he has taught for at least one year was created as a result of the proposal of three field experts whose opinions were consulted. In this context, ten field experts were determined by purposeful sampling. The experience of the field experts participating in the research varies in the range of 5-35 years.

Data Collection Tool

In this study, semi-structured interview forms were used as data collection tool. Semi-structured interview forms consist of three components. The first component of the interview forms was information. The information page contains information on the subject, purpose, scope of the interview. The second component of the interview form is the written consent form. The written consent form includes how to conduct the interview, the recording techniques and tools to be used, by whom under which conditions the data will be stored, how the data will be used for what purpose, sharing the interview data and the contact information of the researchers. In addition, the written consent form includes a two-dimensional written consent table for the participant's participation in the semi-structured interview. The first dimension is for the participant's approval regarding the interview record, and the second dimension is for sharing the quotations directly. The final component of the semi-structured interview form consists of interview questions reflecting the purpose of the research. In terms of opinion and scope validity, the interview questions prepared by applying to the opinions of three academicians who are experts in qualitative research are as follows:

Content generating learners ;

- How can they access correct information in the process of generation content?
- What kind of a path should they follow in order to reach up-to-date information?
- What responsibilities should they have in order to understand that they have reached the correct information?
- In which responsibilities should they reach up-to-date information?

Data Analysis

The transcription and coding of the interview audio recording data obtained from the participants were carried out by the researcher. The data poured from the voice recording to the text was transferred to the NVIVO package program. NVIVO package program was used in the analysis of the data. Descriptive and content analysis was used in the analysis of the data obtained from semi-structured interviews, which is one of the qualitative research methods. Inductive analysis was used in content analysis. Inductive analysis is an analysis method that requires uncovering themes and dimensions, reaching concepts and relationships and analyzing the data in depth (Yıldırım & Şimşek, 2006). For inductive analysis, firstly, the data grouped based on research questions and problems are read, and the nature and general view of the data are presented. Then, based on the research questions, the main themes and sub-themes were extracted from the coded data by coding the data from the relevant text and data sources, and the relationships between them were determined (Creswell, 2013). Finally, the findings were interpreted by enriching them with relevant texts according to their importance in answering the research questions, and interpretation of the data with direct quotations was carried out. Nicknames were used instead of the real names of the participants. The analysis and interpretation of the data obtained from the field experts are given in the findings and comments section.

FINDINGS

The findings obtained by analyzing the semi-structured interview data in accordance with the research purpose and research questions with the NVIVO11 program were supported by direct quotations. Findings obtained from the research are given under the appropriate titles according to the relevant research questions.

Opinions of Field Experts on Accessing the Right Information

From the research questions, "How can content generating learners find the right information during the content generation process?" The answers given by the experts to the question were analyzed with the NVIVO11 package program. As a result of the analyzes, a total of ten sub-themes were determined under the main theme of "Reaching the Right Information". Different opinions, which are connected to two of them, were obtained from the sub-themes reached. Sub-themes collected under the main theme of "Reaching the Right Information"; It was formed by connecting to networks, applying to field experts, reaching scientific information, being included in a learning society, accessing educational resources accessible to access, following corporate pages, guiding learners, using technology, making use of the bibliographies of the resources reached, and using the resources opened by universities. Under the sub-theme of "reaching scientific information", two other views such as "getting out of the information dump" and "scientific studies" were determined. Under the sub-theme of "guiding learners", five such as "enabling access to academic studies", "contributing to learners in academic competence", "comprehending research methods and data collection methods", "adopting discipline rules" and "directing them to relevant field experts" separate opinion has been determined. The sub-themes most emphasized under the Reaching the Right Information main theme are; There were sub-themes of guiding learners and following corporate pages. The opinions obtained under these sub-themes are as follows:

"... We need to encourage learners to read academic publications."

"... Learners do not know how to get the right resources at the first stage. In this regard, the advisors should direct the learners to these resources. "

"... I think that he can get the best resources on the subject that he will generate content through his advisor."

"... Especially if he is researching in electronic environments, the information should be kept up-to-date and reliable, given that it is given by a specific authority."

"... I think of the correct information as information shared with the learners by the educational institution. It is beneficial for a learner to reach the correct information through official channels before generation his own lecture notes and content. "

"... The environments in which they will provide the content should be institutional, formal, that is, they should take care to access the content prepared by the institution. I think they can provide the right information from there. "

Opinions of Field Experts on the Responsibilities of Accessing the Right Information

From the research questions, "What responsibilities should the content generating learners have in order to understand that they have reached the right information?" The answers given by the experts to the question were analyzed. As a result of the analysis, a total of sixteen sub-themes were reached under the main theme of "Responsibility to Access Correct Information". Different views were obtained from five of the sub-themes reached. Sub-themes collected under the main theme of "Responsibility to Access Correct Information"; morality, having a connectivist approach, being an information literacy, reading a lot, critical thinking skills, ethics, seeking feedback, ability to work collaboratively, knowing resource research, not be very lazy, awareness of cultural values, not being biased, self-evaluation skills, self-regulation skills, self-motivation skills, data diversification. Among these sub-themes, the idea of "having critical thinking" is determined under the sub-theme of "being an information literacy". Again under the sub-theme of "critical thinking skills", which is one of the sub-themes; Two different views have been identified, such as "searching the source of information" and "not accepting the information obtained as it is". In this part, the most emphasized sub-themes by the participants were "ethical" and "critical thinking skill". The views expressed under these sub-themes are as follows:

"...You have to be flexible, you have to get information from different people from different cultures, but there must be principles and moral ethical rules about how to filter that information."

"...they need to know the ethical rules first."

"... What we call ethical value actually covers many things. It should have these values in terms of scientific morality. Such responsibilities are very serious. ".

"... Those who content generate should have ethical concerns ..."

"... It turns out that learners should have critical thinking skills here. Analysis, synthesis, evaluation, problem detection, expressing intuition, finding different solutions for this... "

"... Learners must have critical thinking skills to generate original ideas..."

"... somehow it appeared on a newspaper or on a web page, then this information is true, or there is a writer who is constantly being followed, it should not be taken into account that the information she shared is absolutely correct."

Opinions of Field Experts on Accessing Up to Date Information

From the research questions, "What kind of a path should the content generating learners follow in order to reach up-to-date information in the content generation process?" The answers given by the experts to the question were analyzed. As a result of the analysis made with the NVIVO11 package program, main themes and sub-themes were reached from the interviews. A total of thirteen sub-themes were reached under the main theme of "Reaching Up-to-Date Information". Different opinions, which are connected to two of them, were obtained. Sub-themes collected under the main theme of "Reaching Up-to-Date Information" to apply to field experts, to form instantaneous trends, to pay attention to the context, to rapidly advance the knowledge, to be included in a learning society, to follow the counselors through networks, language proficiency, following corporate pages, to use mobile technologies, to benefit from social networks, to follow periodicals, using technology effectively, publication date. Under these sub-themes, the theme of "follow the counselors through networks" was determined as "a different view from the advisors". Again, under the theme of "follow corporate pages", which is one of the sub-themes, the view of "responsibility of institutions" was determined. The most emphasized themes under the main theme of Reaching Up-to-Date Information were "following corporate pages" and "publication date". The opinions obtained under these themes are as follows:

"...For up-to-date information, I think that the content of the activities or announcements of the institution or related unit should be taken as basis."

"...There are also corporate pages or a number of corporate associations, for example, it is necessary to follow their social network pages such as IT association."

"...maybe they should be able to access professional networks in online sharing environments.".

"...In order to reach up-to-date information, the learner must first look at the publication date of the information he / she has received in electronic media."

"...The biggest criterion of this is, of course, the date of publication and the fact that the information is based on the latest events, the latest developments, the latest technologies and references regardless of the subject, and if the information is based on up-to-date events, this information can be estimated to be up to date."

Opinions of Field Experts on the Responsibilities of Accessing the Up-to-Date Information

Within the scope of the study, "In what responsibilities should learners have access to updated information? The answers given to the question were analyzed through the NVIVO11 package program. As a result of the analyzes, a total of fourteen themes were reached under the main theme of "responsibility to access up-to-date information". Different opinions, which are connected to two of them, were obtained. Sub-themes gathered under the main theme of "Responsibility to Access Up-to-Date Information"; multi-tasking, seeking ethics, feedback, turning to an interesting topic, keeping up with the global age, being up to date learners, learning responsibility, not being biased, having self-evaluation skills, having self-regulation skills, self-motivation skills possessing self-direction skills, taking copyright, diversifying data. Among these sub-themes, the view of "being competitive and courageous" was determined under the theme of "keeping up with the global age". Under the sub-theme of "being up-to-date learners", another sub-theme, the view of "being active" was reached.

Apart from the answers given to the research questions determined within the scope of the study, it was also asked whether there is any information that they would like to add to the field experts during the interview. The answers of the field experts whose opinions were taken as additional information were analyzed. As a result of the analysis, a total of ten sub-themes were reached under the main theme of "content generation". Another view related to one of the sub-themes reached was obtained. Sub-themes collected under the main theme of "Content Generation "; accurate information, everyone's own content generation, quality control process, cultural elements, media literacy skills, roles in the learning environment, sharing, a painful process, different perception of the generating content, and Prosumer. Depending on the theme of "sharing" from these sub-themes, the "control of the content generating" view was also obtained. The sub-themes highlighted in this section are "being up to date learners" and "keeping up with the global age" sub-themes. The opinions obtained under these sub-themes are as follows:

"...is the learner up to date? He needs to devote his time to it so he can improve himself."

"...In order to reach up-to-date information, learners must also follow current events."

"...What were the trends a few years ago, now what? Where does the content go when we say content generation? In order for the learners to be up to date, they need to read them a little."

"...he must have been self-actualized. What is this? There must be people who know what they want and do for what they want, who can be aware of their competencies and keep up with the global age."

"...Now we are in the global age, now the borders have been lifted and we are now competing not only with the people in our own country but also with global people. Therefore, the more we get the up-to-date information from the dynamic market where information flows, the more step we will stand out."

General Findings

As a result of the analyzes carried out within the scope of this research, a total of five main themes were obtained. The first of these main themes is the "reaching the right information" main theme. The main theme of "reaching the right information" states what the content generating learners should have and what should be considered in order to be accurate in the information they will obtain and use in the content generation processes. Based on the opinions of the field experts, how the learners can access the correct information is expressed under this main theme. The sub-themes that received the most reference for the learners to reach the correct information were the sub-themes of "guiding the learners", "following the corporate pages", "reaching the scientific information". In this context, it can be said that the information that the learners should accept as true information should be scientific information. In addition, the provision of information by an authority such as advisors or official institutions can be seen as a necessary point in accessing correct information.

The second main theme obtained was the main theme of "Responsibility to access correct information". The main theme of "Responsibility to access correct information" addresses the attitudes and behaviors that content generating learners should have in obtaining the correct information. At the point of reaching the right information, it is important to assume the responsibilities specified by the field experts by the learners and instill the distance education institutions to the learners in order to generate the content containing the correct information. Sub-themes that receive the most references under the responsibility of accessing correct information; There were sub-themes of "ethics", "critical thinking skills", "data diversification", "asking for feedback", "ability to work collaboratively". Therefore, it can be said that learners should firstly be responsible for being ethical. In this context, it is a few of the responsibilities that learners should be able to distinguish between right and wrong and references of information obtained. In addition, it is the important responsibilities that learners should undertake to cooperate with other peers, to pass the information obtained through the mind filter, to ask for feedback from field experts or consultants, and to determine the accuracy of the information obtained from more than one source.

As a result of the analyzes made within the scope of the research, the main theme of "reaching up to date information" was determined as the third main theme. In line with the main theme of "reaching up to date information", the features required to capture the diary of the learners in the content generate process and to use this up-to-date information are mentioned. Based on the opinions of the field experts, it is stated under this main theme how the learners can access up-to-date information and what features they should have. The most repeated sub-themes in the content generation process for the learners to access up-to-date information were: "Following Corporate Pages", "Publishing Date", "Following Periodicals" and "Using Social Networks". Based on the sub-themes obtained, it is seen that the content generating learners should follow the corporate pages at the point of reaching the correct information as well as at the point of reaching the up-to-date information. In addition, the content generating learners look at the history of the information they have obtained, obtain this information from periodicals and use social networks.

The fourth main theme obtained in the research was the "responsibility to access up-to-date information" main theme. The main theme of "responsibility to access up-to-date information" refers to the responsibilities that content generating learners should take on at the point of catching up to date. It can be seen as an important point in terms of the fact that the responsibilities specified by the field experts are assumed by the content generating learners and the educational institutions to which the learners are affiliated also overcome these responsibilities to them.

Sub-themes that received the most reference on the responsibilities that learners should have in reaching up-to-date information were "having self-assessment skills", "being up-to-date learners", "keeping up with the global age", and "ethical" sub-themes. At this point, it can be said that learners should have ethical concerns again. In addition, fulfilling the requirements of the era they live in and being able to evaluate themselves are seen as a necessary element for the learners to access up-to-date information.

The last main theme obtained is the main theme of "content generation". Apart from the research questions, the main theme of "content generation" was determined in line with the opinions expressed by the field experts. Within the scope of this main theme, what other concepts should be considered in the process of content generation and what other features should be considered. Sub-themes that received the most reference in the content generation process were the sub-themes of "roles in learning environments", "quality control process", "different perception of generating content", "correct information" and "cultural elements". Based on the sub-themes obtained, it can be said that there should be a quality control process in the process of generation content. It is also seen that the roles in the learning environment need to be restructured in the content generate process. In addition, the fact that the learners can define the correct information and pay attention to the cultural factors, does not lead to the perception of the content generating differently than they are among the topics discussed.

CONCLUSION, DISCUSSION AND SUGGESTIONS

In this study, it is aimed to determine the characteristics of learners who are registered to a course given in digital learning environments through distance education and generate the content they need in their learning processes. It is possible to say that the roles in learning processes have changed with the development of technology and its inclusion in learning environments. With the changing roles, it is seen that the learners have become producers rather than consumers. Learners who take responsibility in learning environments create learning content in line with their own learning needs. The content generating learners in the related literature are defined as individuals who carry out learning activities by interacting with each other in line with a specific target and generate learning content in this process (Boettcher, 2006). In order for the learners involved in digital learning environments to generate content, they must be able to fulfill the requirements of the environments they are connected to. In other words, learners should have 21st century skills such as technology literacy and media literacy in digital learning environments where technology is intensely employed (Leonard et al., 2007). It can also be said that learners with 21st century skills should also have a connectivist approach in order to be able to interact and communicate with each other or with instructor.

Connectivism explains learning through networks in the 21st century digital information age (Siemens, 2004). In this context, the fact that the learners, who are included in digital learning environments, have both 21st century skills and a connectivist approach, will increase the efficiency of the learning processes.

The results reached within the scope of the study are about how the content generating learners can select and use the correct and up-to-date information from the stack of information they are exposed to during their content generation processes. The findings obtained as a result of the analysis made based on the opinions of the field experts are expressed as the main theme and sub-theme. In line with the findings obtained, it becomes an important requirement for the content generating learners to reach the correct information, to follow the corporate pages and to provide the information by an authority. At this point, the fact that the learners know the extensions of the websites they access will be effective on whether or not they access the corporate page. Learners need to know internet extensions such as .gov (government institutions), .edu (educational institutions), .org (non-commercial institutions), .mil (military institutions), .net (service providers), .int (international institutions). Thus, it can be determined whether the resources they reach are institutional and reliable. In addition, the fact that learners can connect to networks and follow relevant field experts over networks is seen as an important result in reaching correct information. Therefore, it can be said that the learners should make sure that this information comes from reliable sources before using any information they access. Along with these, it was another result that learners should be able to use consultants in content generation processes. At this point, it is an important point for the instructor to implement the role of guidance in learning processes well, so that the learners can generate valid and reliable content. It is thought that instructor should know the digital learning environments they are involved in in order to be a good guide. In parallel with this view, Guasch et al. (2010) argues that instructor should be able to use the communication systems that will be used in digital learning environment. Within the scope of the study, the guidance of the instructor consists of six steps. These steps are:

1. To allow access to academic studies,
2. Contributing to learners on academic competence,
3. To comprehend research methods and data collection methods,
4. Transferring print or online data collection resources,
5. Adopting academic discipline rules,
6. Directing to relevant field experts

The fact that learners know that they need to master the subject and where to learn about it is an important point for the generation of valid and reliable content. One of the important requirements that arise at this point is to become an information literacy. There are two different approaches to the definition of information literacy. While the first one focuses on the definition, the other focuses on the characteristics that individuals should have. In the first, Taylor (1979) interpreted information literacy as accessing information in order to solve real-life problems, accessing information from different sources, ensuring the continuity of information, and knowing the strategies about when and how to obtain information. In the second approach, Doyle (1994) listed the characteristics of individuals who are information literacy as follows: Knows that correct and sufficient information is the basis for making reasonable decisions, realizes the need for information, formulates questions based on information needs, determines possible information sources, develops successful research strategies, accesses information sources including computer-based and other technologies, evaluates information, practical application, organizes information for its transformation, integrates new information with existing information, uses information in critical thinking and problem solving.

Based on the findings obtained in reaching the correct information, another important requirement arises from the fact that the learner is a technology literacy and can be included in the learning communities by connecting to the networks. In addition, the fact that the learners have access to scientific publications with the guidance of instructor has been an important inference in reaching the right information.

In the responsibility of reaching the right information, the issues that learners should think critically and ethically concern are highlighted. Regarding critical thinking, Paul and Elder (2006) stated that there is a positive relationship between quality of life and quality of thinking. Therefore, critical thinking is an important feature that takes place in every moment of life, such as the ability of learners to better adapt to a complex and constantly changing environment and to improve the quality of life (Torres and Cano, 1995). Information that does not pass through the critical thinking system has a biased, distorted, biased, uninformed and reductive feature (Paul and Elder, 2006). İpşiroğlu (1997, p.33) defines critical thinking as follows. "It is a way of thinking that goes down to the essence of problems, examines from different perspectives, tries to understand, and can oppose if necessary."

Based on the findings of data diversification, it is concluded that learners should research the information they have accessed from more than one source and provide information in this way. Another result is that the learners not be lazy in their learning processes and should have self-regulation and self-evaluation skills instead of being content with the knowledge of others. In line with the finding of guiding the learners about reaching the correct information, the finding that the learners want feedback is reached. Therefore, it can be concluded that the learners should have the responsibility to ask for feedback first. The request of the learners for feedback can be seen as an element that can cause instructor to guide. Other responsibilities arise that learners should have a connectivist approach and work collaboratively. At this point, it can be concluded that learners who have a connectivist approach can cooperate with other peers over the networks without having any physical limitations and can help in learning processes. Based on these results, it can be said that the needs in digital learning environments are more diversified than those in traditional learning environments. In parallel with this view, Aldemir (2003, p.272);

"The development of literacy based on human abilities and ingenuity caused some issues to come to the agenda in the 21st century. We can list these topics as follows: realizing the need for information, locating information, accessing information, creating search strategies, analyzing, interpreting, evaluating, communicating, having problem solving skills, resource-based education, lifelong learning, learning society, continuous education, education curriculum reforms and respect for copyrights." has used the expression. Therefore, it can be said that the learners should renew themselves with the innovations that result from the development of technology.

It has been a frequently repeated finding by field experts that content generating learners should follow periodicals to access up-to-date information. In this context, it can be concluded that academic publications that make periodicals in content generation processes are important in terms of providing updated information. Also, looking at the publication date of the content reached is another finding. It has been determined as an element to be underlined that the learners should check the content they have reached, especially in terms of catching up to date, especially their publication dates. The finding of applying to field experts expresses that learners should be able to connect to networks. It can be said that learners should have skills such as technology literacy, media literacy, which are 21st century skills, in order to connect to networks. In this context, it can be concluded that the 21st century skills theory and connectivism theory, which are handled within the context of content generating learners, are important in terms of generation content in learning processes. Based on the findings obtained, the use of mobile technologies by learners has also been an important result in terms of following other peers, instructor and field experts over networks. It can be said that it is advantageous in terms of obtaining instant information for learners who are included in a learning environment over networks. Again at this point, it is possible to say that 21st century skills and connectivist approach must be employed. In parallel with this view, it is thought that learners with 21st century skills and connectivist approach can access instant information by making efficient use of social networks. It can be said that one of the most important issues that learners should focus on in reaching up-to-date information is the context of the content to be generating. If the context of the subject that the content generating learners in the learning processes does not require to be up to date, the learners may not worry about catching up to date. For example, if a historical process is considered as the subject studied, it can be concluded that the learners should turn to the right information at this point. In addition, the fact that learners follow instant trends in environments where information is rapidly advancing is an essential element in accessing up-to-date information. It is possible for learners to reach instant trends on networks with the help of mobile technologies. Therefore, it is also possible for learners who can use mobile technologies to access instant information on networks.

The most important result obtained from the findings about the responsibility of the content generating learners to access up-to-date information was that the learners should have self-evaluation skills. In this context, it can be said that in order to reach the up-to-date information, the learners should have a need or have a responsibility for learning in the content generate processes. It can be said that self-assessing learners can identify the needs they need and generate content accordingly. Another result obtained in the scope of the study was that in order for the learners to have access to updated information, the learners should be able to keep up with the requirements of the age they live in, in other words, to be up to date. The elements necessary for the learner to stay up to date are determined as follows:

1. Attending conferences,
2. Using mobile technologies,
3. Using social media accounts,
4. To make readings in the related field,
5. Being active,
6. Interdisciplinary work,

Another factor emphasized by the field experts was that the learners were not biased. From this point of view, it is concluded that valid, reliable or up-to-date information in sources that learners do not care about can be ignored.

Within the scope of this study, opinions of the field experts other than the research questions determined were also taken. The opinions of the field experts also overlap with the statements stated in the field literature of this study. One of the findings that emerged as a result of analyzing the opinions obtained was about changing the roles in learning environments. In order not to lose control of learning environments, the environments where instructional and technical choices are made by instructor and commands are given by instructor, which are one way, are mostly traditional learning environments (Siemens, 2004). However, this model does not mean that the functions that instructor have will disappear over time (Skrypnyk et al., 2015). In the learning processes carried out in networked systems, control among learners can be achieved by guiding learners, supporting and directing learners, and supporting learners socially (Siemens, 2010). Therefore, it is possible to say that the role of instructor has changed in a network-based digital learning environment that has emerged with the development of technology and has evolved from a centralized structure to a structure for guidance. In parallel with this situation, it can be said that learners in digital learning environments are at the center of the learning process, and responsibilities in learning environments are shared with learners. Regarding the roles of learners in the digital learning environment, Richardson and Newby (2006) stated that learners involved in digital learning environments gained experience and assumed more responsibility for their own learning experiences.

Another finding overlapping with the literature was that there should be a quality control mechanism in the content generation process. The field experts whose opinions are consulted within the scope of the study argue that the content prepared by the learners in the process of content generation should be reviewed by the relevant experts or other peers. In parallel with these opinions, Ehlers (2009) and Perez-Mateo et al. (2011) stated that the existence of a quality control mechanism is important in digital learning processes. In this way, more reliable content will be generated.

It was concluded that one of the most important steps in the content generation process is to define the correct information. After defining the correct information, it will be easier for learners to access this information. Based on the opinions of the field experts, it is possible to define the correct information as "meeting the individual learning needs of the learners". Therefore, it is not possible to talk about the existence of a single correct information. The political view of the learner, the economic situation, gender and their roles in the society also affect what information they perceive correctly. Learners should try to access this information after defining the correct information in line with their needs. In the learning processes carried out within the scope of a course, the information required by the subject of the course can be perceived as correct information.

Suggestions

Keeping up with the roles in the learning environment is an essential element for contemporary learning environments to provide rich learning outcomes. Sharing learning responsibilities is one of the basic elements in changing learning environments with developing technologies. Therefore, instructor who act as guides in learning environments should give learners responsibility in their learning processes. In this context, it can be said that in digital learning environments where learners can generate content by taking responsibility, it is necessary to have control over synchronous or asynchronous communication systems that instructor must use. In other words, instructor need to be able to keep themselves up-to-date and use the necessary technologies to guide them. In this way, it is thought that it can be beneficial for learners. Similarly, it is an important point for learners to have the requirements and responsibilities achieved within the scope of this study and to generate valid and reliable content.

Ensuring that the content generating by the learners is shared on digital media at the end of the content generation process is seen as an element that will improve the technology literacy of the learners (Bakla, 2018; Song et al., 2019). At the same time, learners who share their products will be able to develop their cooperation skills and gain self-confidence thanks to their peers who give feedback or write comments to them under their share. In this context, ensuring the sharing of contents is an important factor.

As mentioned earlier, digital learning environments using Web 2.0 tools make positive contributions in terms of achieving targeted learning outcomes. Therefore, the use of such learning environments in institutions that provide open and distance learning services can be seen as an important point in terms of learners' academic success, collaborative work skills and connectivism. In this way, it is possible to prevent learners from having limited time and space.

Based on the opinions of the field experts and the related literature, it is thought that the environments in which the learners generate content will become more widespread in the future. From this point of view, it is considered as a necessary factor for distance education institutions to provide the learners with learning and responsibilities in the learning processes, in order to provide valid and reliable content in the future. Therefore, training of distance education institutions on how to access the correct and up-to-date information will increase the efficiency of digital learning environments. In this way, it is thought that the learners walking on the networks will go through a healthier learning process.

NOTE

The current study utilized the dataset of the first author's master's thesis.

REFERENCES

Aldemir, A. (2003). *Bilgiye erişimde yeni yaklaşım: Bilgi Okuryazarlığı*. Aytaç Yıldızeli, Canan Duran, Hatice Kübra Bahşişoğlu.

Altunışık, R., Coşkun, R., Bayraktaroğlu, S. and Yıldırım, E. (2010). *Sosyal Bilimlerde Araştırma Yöntemleri SPSS Uygulamalı* (6. Baskı). Sakarya: Sakarya Yayıncılık.

Bakla, A. (2018). Learner-generated materials in a flipped pronunciation class: A sequential explanatory mixed-methods study. *Computers & Education*, *125*, 14–38. doi:10.1016/j.compedu.2018.05.017

Belanche, D., Casaló, L. V., Orús, C., & Pérez-Rueda, A. (2020). Developing a Learning Network on YouTube: Analysis of Student Satisfaction with a Learner-Generated Content Activity. In *Educational Networking* (pp. 195–231). Springer. doi:10.1007/978-3-030-29973-6_6

Creswell, J. W. (2005). *Educational Research: Planning, Conducting, and Evaluating Qualitative and Quantitative Research* (2nd ed.). Pearson Education.

Creswell, J. W. (2007). *Qualitative inquiry and research method: Choosing among five approaches.* Academic Press.

Creswell, J. W. (2013). Research design: Qualitative, quantitative, and mixed methods approaches. *Sage (Atlanta, Ga.).*

Dale, C., & Povey, G. (2009). An evaluation of learner-generated content and podcasting. *Journal of Hospitality, Leisure, Sport and Tourism Education*, *8*(1), 117–123. doi:10.3794/johlste.81.214

Doyle, C. S. (1994). *Information literacy in an information society: A concept for the information age*. Syracuse University.

Fırat, M., Yurdakul, I. K. & Ersoy, A. (2014). Bir eğitim teknolojisi araştırmasına dayalı olarak karma yöntem araştırması deneyimi. *Eğitimde Nitel Araştırmalar Dergisi, 2*(1).

Fischer, G., & Konomi, S. (2005). Innovative media in support of distributed intelligence and lifelong learning. In *Proceedings of the Third IEEE International Workshop on Wireless and Mobile Technologies in Education*, (pp. 3-10). Los Alamitos, CA: IEEE Computer Society. 10.1109/WMTE.2005.35

Guasch, T., Alvarez, I., & Espasa, A. (2010). University teacher competencies in a virtual teaching / learning environment: Analysis of a teacher training experience. *Teaching and Teacher Education*, *26*(2), 199–206. doi:10.1016/j.tate.2009.02.018

İpşiroğlu, Z. (1997). *Eğitimde yeni arayışlar*. Adam Yayınları.

Kılınç, H. (2016). *İçerik Üreten Öğrenenlerin Özelliklerinin Değerlendirilmesi. Yüksek Lisans Tezi* (Master Thesis). Anadolu Üniversitesi, Sosyal Bilimler Enstitüsü, Eskişehir.

Klamma, R., Cao, Y., & Spaniol, M. (2007). *Watching the Blogosphere: Knowledge sharing in Web 2.0.* Paper presented at the International Conference on Weblogs and Social Media, Boulder, CO. https://www.icwsm.org/papers/2--Klamma-Cao-Spaniol.pdf

Lambert, C., Philp, J., & Nakamura, S. (2017). Learner-generated content and engagement in second language task performance. *Language Teaching Research*, *21*(6), 665–680. doi:10.1177/1362168816683559

Lambert, C., & Zhang, G. (2019). Engagement in the Use of English and Chinese as Foreign Languages: The Role of Learner-Generated Content in Instructional Task Design. *Modern Language Journal*, *103*(2), 391–411. doi:10.1111/modl.12560

Lee, M. J. W., & McLoughlin, C. (2007). Teaching and Learning in the Web 2.0 Era: Empowering Students through Learner-Generated Content. *International Journal of Instructional Technology and Distance Learning*, *4*(10).

Lee, M. J. W, McLoughlin, C. & Chan, A. (2008). Talk the talk: Learner-generated podcasts as catalysts for knowledge creation. *British Journal of Educational Technology, 39*(3), 501–521. DOİ: doi:10.1111/j.1467-8535.2007.00746.x

Lenhart, A., & Madden, M. (2005). *Teen content creators and consumers*. Pew Internet & American Life Project.

Leonard, A. A., Elizabeth, F., & Marta, K. (2007). V-Learning Distance Education in the 21st Century Through 3D Virtual Learning Environments. Academic Press.

Luckin, R. (2006). Understanding learning contexts as ecologies of resources: From the zone of proximal development to learner generated contexts. *World Conference on E-Learning in Corporate, Government, Healthcare, and Higher Education, 2006*(1), 2195-2202.

Luckin, R., du Boulay, B., Smith, H., Underwood, J., Fitzpatrick, G., Holmberg, J., Kerawalla, L., Tunley, H., Brewster, D., & Pearce, D. (2005). Using Mobile Technology to Create Flexible Learning Contexts. *Journal of Interactive Media in Education*, *2005*(2), 21. doi:10.5334/2005-22

Narayan, V. (2011). *Learner-generated content as a pedagogical change agent*. Ascilite.

O'Reilly, T. (2005). *What is Web 2.0: Design patterns and business models for the next generation of software*. https://www.oreilly.com/pub/a/Web2/archive/what-is-Web-20.html

Paul, R. & Elder, L. (2006). *Minik eleştirel düşünme kılavuzu: Kavramlar ve araçlar* (M. Bektaş, Çev). Academic Press.

Pérez-Mateo, M., Maina, M. F., Romero, M., & Guitert, M. (2011). Learner Generated Content: quality from students' point of view. *World Conference on Educational Multimedia, Hypermedia and Telecommunications*, 2011(1), 2520-2529.

QMPP Project. (2009). *QMPP Handbook Quality Management of Peer Production*. Author.

Richardson, J. C., & Newby, T. (2006). The Role of Students' Cognitive Engagement in Online Learning. *American Journal of Distance Education*, *20*(1), 23–37. doi:10.120715389286ajde2001_3

Sener, J. (2007). In search of student-generated content in online education. *E-Mentor, 21*, 1-8.

Siemens, G. (2004). Connectivism: A learning theory for the digital age. International. *Journal of Instructional Technology and Distance Learning*, *2*(1), 3–10.

Siemens, G. (2010). *Teaching in Social and Technological Networks*. Connectivism.

Skrypnyk, O., Joksimović, S., Kovanović, V., Gašević, D., & Dawson, S. (2015). Roles of course facilitators, learners, and technology in the flow of information of a cMOOC. *The International Review of Research in Open and Distributed Learning*, *16*(3). Advance online publication. doi:10.19173/irrodl.v16i3.2170

Song, Y., Lung, K. M., Ogata, H., Yang, Y., Hasnine, N., & Mouri, K. (2019). Bridging in-class and real life vocabulary learning using a mobile learner-generated content tool: A case study. In *Teacher forum proceedings of the 23rd Global Chinese Conference on Computers in Education (GCCCE2019)* (pp. 176-186). Central China Normal University.

Taylor, C. (1979). What's wrong with negative liberty? *The Idea of Freedom,182.*

Toffler, A. (1996). Third Wave (3rd ed.). Istanbul: Altin Kitaplar.

Torres, R. M., & Cano, J. (1995). Learning style: A factor to critical thinking? *Journal of Agricultural Education*, *36*(4), 55–62. doi:10.5032/jae.1995.04055

Türnüklü, A. (2000). *Eğitimbilim Araştırmalarında Etkin Olarak Kullanılabilecek Nitel Bir Araştırma Tekniği: Görüşme. Kuram ve Uygulamada Eğitim Yönetimi Dergisi. 24.* PegemA Yayıncılık.

Yıldırım, A. and Şimşek, H. (2005). *Sosyal bilimlerde nitel araştırma yöntemleri.* Ankara: Seçkin yayınları.

Yıldırım, A., & Şimşek, H. (2006). *Sosyal bilimlerde nitel arastirma yöntemleri.* Seçkin Yayıncılık.

Yıldırım, A., & Şimşek, H. (2008). Sosyal bilimlerde nitel araştırma (6. Baskı). Ankara: Seçkin Yayınları.

Yin, R. K. (1984). *Case Study Research: Design and Methods*. Sage Publications.

Yin, R. K. (2002). *Case study research: Design and methods* (3rd ed.). Sage.

Yin, R. K. (2003). *Case study research: Design and methods* (3rd ed.). Sage.

Yin, R. K. (2009). *Case study research: design and methods. essential guide to qualitative methods in organizational research.* Academic Press.

Chapter 9
Design of Communication Environments in Career-Focused Learning Communities

Buket Kip Kayabaş
Anadolu University, Turkey

ABSTRACT

Rapid renewal of information with the rapid development of technology prevents individuals from using their knowledge and skills throughout their lives. Today, instead of participating in long education programs such as undergraduate or associate degree programs, short-term education programs that are specialized according to their fields of study are increasingly in demand. Investigating which communication environments individuals prefer in career focused learning communities in which individuals participate in lifelong learning is important in terms of shedding light on the design process of these environments. The purpose of this manuscript is firstly to reveal the opportunities offered by lifelong learning to individuals, then to determine the preferences of the communication medium based on the experiences of individuals participating in career focused learning communities offered within the scope of lifelong learning, and finally, to provide suggestions for the design of communication environments to people who create and manage career focused learning communities.

DOI: 10.4018/978-1-7998-4534-8.ch009

INTRODUCTION

Today, where the information is produced, accessed, shared and consumed rapidly with the change of information and communication technologies, there is a need for individuals to catch up with the changes and update themselves constantly in order to be successful in social and professional lives. After losing the acquired information through diploma programs that are outdated within 3-5 years, individuals turn themselves to educational programs where they can always keep themselves up to date. While they are actualizing this, they are also supposed to carry the responsibilities of an adult life. Through open and distance learning, career focused learning communities emerging from this need provide individuals opportunities to progress at their own pace of learning without leaving their daily life behind and to have flexible experiences that are open to communication and interaction. Career focused lifelong learning communities can be defined as communities formed by learners who want to improve themselves within the scope of professional, individual, and social skills. In keeping up to date professional, individual, and social skills in today's business world; lifelong learning communities such as e-certification programs, private courses and MOOCs have gained great importance.

In career focused learning communities conducted through open and distance learning, continuity of lifelong learning is ensured by providing individuals with different opportunities and features working in different works that are having opportunities to have an access to the education in desired fields and improve themselves. Different communication opportunities in these communities, such as synchronous and asynchronous communication, are offered to provide social interaction. Individuals communicate with both employees of the institution and/or other learners by choosing the communication environment that suits their habits and preferences. Conducting studies to determine the preferences of the communication environment of the learners and to investigate what should be considered before designing the courses is important to shed light on these aspects for the instruction designers.

The main purpose of this manuscript is to determine the communication environment preferences of individuals participating in career-based learning communities offered within the scope of lifelong learning. For this purpose, the answers to the following questions were sought:

- Who participated in the programs?
- What are the participants' experiences in career focused learning communities?
- What are the main motivations of participants to participate in career focused learning?

- What are the communication preferences of the participants in career focused learning communities?
- What are the learners' views on career focused learning communities?

Literature Review

Even though it varies according to the needs of countries and education policies, lifelong learning practices in general; envisages to ensure that individuals become active citizens in the society by taking into account the social, cultural and professional development of individuals, including all forms of learning approaches that are structured, unstructured, semi-structured for all learners. In literature, lifelong learning is defined in different ways such as a personal, social and professional development process to improve the quality of life of individuals and others who live with them (Dave, 1976, p. 12), providing educational opportunities to the individual throughout his life (Kulich, 1982, p. 123), accessing the information they need to help individuals perform and manage their lives (White, 1982, p. 75), personal planned, unplanned, experience-based development process (Holmes, 2002, p. 2), all learning activities that are structured, unstructured or semi-structured to improve the knowledge, skills and abilities of individuals (European Commission, 2002), a combination of individual and institutional learnings (Jarvis, 2004, p. 130). In summary, lifelong learning can be described as a learning process that continues throughout one's whole life in order to reach the rapidly developing, changing and renewed information that will help the individual to succeed in social and professional life, cope with the problems, provide solutions when needed, and develop his competencies. Lifelong learning is accepted as a basic policy by the European Parliament for the development of active citizenship, social cohesion and employment, along with its being emphasized that it should be perceived as a process that should spread throughout every stage of life rather than necessity (Gündoğan, 2003, p. 3).

The goals of lifelong learning are expressed as follows (European Commission, 2002):

- Considering the needs of individuals, to structure an education system in which they can reach qualified education throughout their lives with equal opportunities for everyone in the society,
- To ensure the adaptation of individuals and social integration to innovations that change rapidly in professional knowledge, skills, and methods,
- To encourage the active participation of individuals in all areas of social, cultural and political life and to equip them with the necessary competencies.

It is emphasized that in lifelong learning process an approach in which the individual is at the center is adopted, the role of the school is changed and nonformal learning is gained importance, the control of the state on the learning processes is reduced, social roles are strengthened and learning experiences are not limited in a certain given period of time (Güleç et al., 2012, p. 34; Wain, 1993, p. 58). Lifelong learning as a concept was first expressed as "learning from birth to death" at a conference organized by UNESCO in 1960 (European Commission, 2002). Reports were published at the end of the 1990s emphasizing that lifelong learning has a strategic importance in adapting to new technologies, social and economic developments (European Commission, 1994; European Commission, 1995; European Commission, 1996). Then, when the core-decision procedure started in the European Union, 1996 was declared as the "European Year of Lifelong Learning". According to that, the aims are to increase the quality of teaching activities, strengthen qualifications of young people with vocational education, encourage individuals to improve, and increase the awareness of the parties and parents about lifelong learning by developing cooperation between the business world and educational institutions (Toprak and Erdoğan, 2012, p. 77). During the Lisbon Summit held in 2000, "Lifelong Learning Memorandum" was issued, emphasizing the need to focus on lifelong learning within the basic employment policies of the European Union (Gündoğan, 2003, p. 5). According to this memorandum, six key principles were determined for the lifelong learning strategy to be adopted by member states in the 21st century, by stating that lifetime learning should be emphasized to promote active citizenship and employment (CEDEFOP, 2001):

- Universal and continuous education should be guaranteed.
- The investment in human resources should be increased.
- Innovative teaching methods should be applied.
- Better and more clear methods for the assessment and evaluation of the learning process should be developed.
- An access of guidance and counselling should be built.
- Provide access to everyone for lifelong learning opportunities with the help of new technologies.

Then, "Bologna Seminar on Lifelong Learning" was held in Prague (BFUG, 2003) in 2003, and the topics such as recognition of prior learning, organizing more flexible learning opportunities, and widening access to higher education were brought to the agenda (Toprak and Erdoğan, 2012, p. 73). The 2007 London Communique (BFUG, 2009) and the seminars that held in 2008 have also emphasized that individuals could improve themselves with the help of lifelong learning, could be employed by keeping up with changing social and economic conditions, and become active citizens.

After briefly mentioning the history and philosophy of lifelong learning, its features, scope and possibilities can be mentioned. Lifelong learning covers a wide and various range of processes from structured learning to unstructured experiences (CERI, 2001; OECD, 2004). On the other hand, the learner is at the forefront with his needs, demands, learning styles and individual characteristics in the lifelong learning process. Individuals will experience learning in their own motivation and pace throughout their lives. Therefore, individuals who have access to the economic, cultural and social gains with the help of lifelong learning opportunities can become more participatory and active citizens in the society. On Table 1, the differences in approach between traditional learning experiences and lifelong learning opportunities are summarized (Toprak and Erdoğan, 2012, p.73). By examining Table 1, the source of information is seen to be instructive in institutions where traditional learning experience takes place. The process of creating, sharing, and consuming information is happening very fast, especially with the developments in information and communication technologies. Thus, it can be said that the instructors should become a guide to the learners about how to access the information that they will need, not just during the school years but also throughout their lives. On the other hand, it should be taken into consideration that learning experiences are lifelong processes rather than being acquired at a certain age in schools, and that individuals will learn together with other individuals in the social group they live in. While traditional learning takes place in a frame of transferring information from instructor to individuals in an institution, lifelong learning focuses on the development of individual abilities and learning capacities. The basics of lifelong learning involve the approach of teaching to learn (European Commission, 2002). Individuals can keep the learning process active by accessing, organizing, and reflecting the information they will need throughout their life independently from any institution and time.

The economic, social and technological changes and developments; and the level of knowledge and skills that individuals should increase rapidly and simultaneously (European Commission, 2002). Basic competences are communication skills in mother tongue, mathematical competence and basic competence in science and technology, learning to learn, social and civic competences, cultural awareness and expression and digital competence (European Commission, 2002; European Commission, 2006). The ability to use information and specifically communication technologies is stated one of the factors affecting the participation of individuals in lifelong learning practices (Tan and Morris, 2006, p. 316). Career focused learning communities that emerged with the purpose of contribution to the vocational development of individuals provide individuals opportunities to progress at their own learning pace, reflect their professional experience, and have flexible learning experiences with open and distance education. Career focused learning communities are the platforms provided by open and distance learning, where open course materials and mass open online courses (MOOCs) can be found. MOOCs, which have their originals in the movement of openness; it can be seen that the courses in this structure are career focused learning platforms that are combined with experiences seem to be independent from each other but serve a common purpose in general. Mass courses are linked tightly to open learning resources and the technologies, pedagogical approaches and methods that used for them are not new; along with this, the experiences gained in the field of open and distance learning and the offered opportunities by information and communication technologies made all of these easily accessible and usable. MOOCs are online courses that are generally free, designed for large groups of students, accessible by anyone with an internet connection, and without any entry requirements. These bring career focused learning platforms under the light to be discussed.

Table 1. Comparison of traditional learning and the experiences of lifelong learning.

Traditional Learning	Lifelong Learning
Instructors are the source of information.	Instructors are in the role of guide for the information sources.
Learners take the information from the instructor.	People learn by doing and experiencing.
Learners study by themselves.	People learn in social settings, with groups and from each other.
Learners are guided and tested until they complete their skills, and are not allowed to move on to the next step.	Assessment is carried to determine the learning strategies and styles.
All learners go through the same process.	Instructors have a role of facilitator and support the individualized learning.
Instructors are trained priorly and then supported by in-service training.	Instructors are also lifelong learners. Prior education is connected to the ongoing vocational education.
Good learners are chosen for the next phase.	All individuals in the society have equal access to lifelong learning opportunities.

Source: World Bank, 2003, p. 29.

Udemy platform, with 17 million students, 20 thousand instructors and 55 thousand courses worldwide, was developed in 2010. Members of this platform can follow the courses that they are interested in by purchasing them and anytime they want. Thus, they have a right of access to these courses for a lifetime. On the other hand, instructors also can upload the courses they have prepared and give access to the students in exchange of a fee. Udacity, Coursera and edX as MOOCs providers were created in 2012. edX (https://www.edX.org) is a non-profit organization that was created by MIT and Harvard University. Anyone can use edX which is an open-source MOOCs platform to create their own learning environment and pursue their courses. Coursera (https://www.coursera.org) is an online learning platform founded in 2012 for profit by New Enterprise Associate, Kleiner, Perkins, Caufield and Byers Education.

Udacity (https://www.udacity.com) is another MOOCs founded by Sebastian Thrun, David Stavens, and Mike Sokolsky. Students who finish one of the courses on Udacity MOOCs platform receive a certificate of completion, signed by the instructors, indicating their level of achievement. Some of the universities offer opportunities to transfer the credits of courses provided by Udacity and its partnership with Pearson exam centre (Yuan and Powell, 2013). FutureLearn (https://www.futurelearn.com) as a special MOOC platform was founded in 2013 with the ownership of Open University and it offers open courses with the partnership of the universities in Europe and private institutions (Yuan and Powell, 2013, p. 7). OpenupEd is an initiative launched in 2013 by the European Association of Distance Teaching Universities (EADTU) to support MOOCs in Europe. OpenupEd is not a MOOCs platform, but it promises to bring some distinctive features to the MOOC landscape.

When the MOOCs are examined in the context of Turkey, a variety of studies can be found. Koç University became a member of Coursera MOOCs platform in 2013 and collaborated with the translation of the three courses to make them available with Turkish subtitles on the platform. Another MOOCs implementation has emerged again in 2013 with the name of University Plus. In 2014, this platform was sold to a different company and it has started to offer paid certification; and then University Plus has been converted from a MOOCs platform to a certificate program with courses offered by Yeditepe and Istanbul Technical University. In July 2014, Turkey's first institutional MOOCs platform AKADEMA was announced by Anadolu University. Initial plan was to announce 50 courses at first and in January 2015, 4 courses went public. Another institutional implementation was AtademiX by Ataturk University. AtademiX was announced in December 2014 and in January 2015 three courses went public. In 2015, Turkcell Digital Academy was announced by Turkcell communications and technology company by establishing a strategic collaboration with MIT and offered Innovator's MBA both in Turkish and English at edX. The platform, created by CampusOnline, offers paid certificate courses by contracted universities. Another institutional attempt was made by Sakarya University. In the same year, Sakarya University collaborated with HarvardX to conduct common courses. In 2016, Middle East Technical University Bilgeİş as MOOCs platform was developed and announced 10 courses.

After the examples of the main career focused learning communities, the research topics, such as which communication medium individuals prefer in these communities for academic, social and managerial interaction and their motivations for signing-up in these communities, should be mentioned. First, it will be useful to examine the reasons for individuals for signing-up in career focused learning communities. The motivations of learners to signing-up in career focused learning communities vary widely, depending on the characteristics of the courses and the means they offer. The researchers stated that one of the most popular motivations of signing-up is the interest of learners' in the subject area (Kay et al., 2013, p. 761; Seaton et al., p. 58). In a similar study, researchers stressed that the main motivation for learners to participate in career focused learning is the curiosity for the subject and the desire to do their job in a better way (Wang and Baker, 2015, p. 17). In another study the motivations of individuals are investigated on a learning platform created by the University of Edinburgh. According to that, the reasons for signing-up are to experience online education, gain an understanding of the subject matter, get certificates, and invest in their professional development (Yuan and Powell, 2013, p. 11).

When the characteristics of career focused learning communities from the diploma programs, learners can be said to register these communities since there are no prerequisite classes that they need to take prior enrolment (Vail, 2013). It can also be stressed that these courses having a characteristic of easy access with internet connection by anyone can be preferred because of the ability to access learning opportunities without having to go to an institution by individuals with physical disabilities (Wilson et al, 2013, p. 110). It is also stated that these courses are preferred because of them being free of charge and accessible at a small fee for a certification (Chen, Barnett and Stephens, 2013; Wilson et al, 2013). Another motivation to prefer these career focused learning communities is the opportunity of taking a course from an expert, knowledgeable and well-known instructor (Wilson et al, 2013, p. 111). Similarly, Adamopulos (2013, p. 16) expressed that some of the massive open online courses are preferred because they are conducted by prestigious, popular universities or well-known professors.

Individuals prefer to use different communication medium while interacting with other learners and the institution in career focused learning communities that they registered for different motivations depending on their needs and tendencies. Different options of medium bring the limitations and richness along with that environment. Students choose a learning material according to their learning styles and their experiences within the given media (Walther and Burgoon, 1992; Williams and Wilson, 1999). Communication media should be enriched with different options of technology and material to help students choose different learning styles and experiences and use the most suitable environment for themselves, especially when designing lifelong learning activities for large audiences.

In literature, there are models describing the choice of media and technology in the design process of the learning environment (Bates, 2003; Heinich, Molenda,

Russel and Smaldino, 1999; Reiser and Gagne, 1983). These models show that the selection of media is made through four main steps:

- Identifying the characteristics of the media based on the teaching objectives and learning activities,
- Identifying the student characteristics that the media appeals to or ignores,
- Identifying the media of learning whether it is suitable or not,
- Identifying the economic and organizational factors that affect the applicability of the media

This model, which consists of four steps and explains the preferences of the media during the design stage of the learning environment, emphasizes firstly on the selection of the environment based on the teaching objectives and learning activities. For example, in a foreign language course, visual and audio technologies should be included in the learning environment. In the second step, the characteristics of students in the choice of media is stressed. If the students are poor in reading, offering audio and video environments will be more appropriate. In other words, while audio environments are offered to students who are visually impaired or who travel a lot; the students with low motivation should be offered more dynamic and interactive environments. Based on the model, in the third step the compatibility of the media with the learning environment is expressed. While some medium are more suitable when reached from home or office due to their high internet bandwidth, some of them can be reached from mobile devices during the journey because of being portable. In the last step, economic and organizational factors that will be needed in developing the medium are mentioned. In this context, the focus is on the budget of the institutions on the accessibility of developers who are experts in this field and the usability of previous experiences.

Communication and learning materials in many distance education programs are designed in a multi-media frame using more than one technology. No single technology is likely to address all the teaching and learning requirements and needs, satisfy different needs of learners or address the variations in their learning environments. Using a mixture of different media together will also allow for differences in student learning styles or capabilities (Moore and Kearsley, 2005, p. 72). Some learners have a reflective thinking and prefer asynchronous environments, while others may prefer live impulsive dialog like virtual classroom activities instead of reflecting on the content. In short, in the case of large groups of learners, the more media options available, the more effective the open and distance learning program will be (Moore and Kearsley, 2005, p. 78).

Studies (Fulk et al., 1995; King and Xia, 1997; Walther and Burgoon, 1992; Williams and Wilson, 1999) stated that one of the most important factors affecting the choice of the media is related to the experiences of it. Similarly, King and Xia (1997, 168) emphasized that the choice of media for the individual to perform a certain task is closely related to one's experience rather than the harmony between the media and the task. Individuals with more experience related to a communication medium tend to prefer using this medium in future again. Therefore, it is worth remembering that the experience is very decisive in terms of preferences. Another dimension of experience is the media diversity. People with more experience can use different combinations of media to fulfil their tasks. Browning et al. (2004) determined that the experienced people in technology use more than one medium while performing their tasks.

The individual characteristics are also effective in the decision of the media (Carlson and Zmud, 1999, King and Xia, 1999). For example, not all the learners prefer to use nonverbal elements. While some express themselves better by speaking, some have a timid character and may prefer to communicate in written medium such as email. Communication is also a determinant. For example, email is preferred for task-oriented communications rather than socially oriented communications (Hinds and Kiesler, 1995; Walther et al., 2001). Similarly, synchronous media such as chat in the virtual classroom is also one of the preferred social communications.

Breslow et al. (2013, p. 13) examined the choice of communication media of individuals who participated in a massive course on edX. When asynchronous discussion forums were investigated, 3% of all the learners initiated approximately over 12,000 discussion threads during the course and 100.000 individual discussion topics. In total 52% of the certificate earners were active on the forum with 28% of them asked a question, 41% answered a question and 31% made a comment (Breslow et al, 2013, p. 20). In a similar study, Grainger (2013, p. 36) stated that only 4% of learners participated in the asynchronous discussion forum during the courses offered on FutureLearn. In another study, Champaign et al. (2014, 11) stated that 15% of the participants of a massive course conducted on edX used the asynchronous discussion panel. Newman and Oh (2014) stressed the demographics of the MOOCs participants having bachelor's degrees and the variety of media usage increased as the level of education did. Researchers stated that learners with master's and doctorate degrees use almost all media environments. Kip (2007, 69) investigated the frequency of the usage of different media provided to open and distance learners, and found out that the students prefer email the most among the provided academic support services, such as email, forum and virtual classroom, offered to them; while they prefer live chat in the virtual classroom the least. The reason to prefer email when communicating in academic matters was linked to preferences of keeping the messages permanent in task-based communications.

Purpose and Research Questions

In this manuscript, it is aimed to determine the communication media preferences of participants in career focused learning communities. For this purpose, answers to the following questions are examined.

- Who participated in the programs?
- What are the participants' experiences in career focused learning communities?
- What are the main motivations of participants to participate in career focused learning?
- What are the communication preferences of the participants in career focused learning communities?
- What are the learners' views on career focused learning communities?

METHODOLOGY

For this study, a descriptive research model was used. A questionnaire was prepared and sent to the learners via email to determine the preferences of the participants about the communication environments during the last year in Anadolu University e-Certificate Programs, which are career focused learning systems. "The Individual Preferences in Career Focused Lifelong Learning Communities" survey consists of four parts. In the first part, demographic characteristics of the participants such as age, gender, education level, employment status were questioned. In the second part, there were questions related to the career focused learning experiences of participants. In the third part, the communication media preferences of the learners in career focused learning, synchronous / asynchronous, were examined. In the fourth part, the opinions of learners regarding career focused learning communities were asked. Questionnaire results can be found on the Findings and Discussion section.

The Context of the Research

This study was designed in the context of Anadolu University e-Certificate Programs, a career focused lifelong learning platform. Since 1982 in Turkey, Anadolu University has been providing undergraduate and graduate degrees by using distance education technologies in Open Education system, and e-Certificate Programs on internet (http://e-sertifika.anadolu.edu.tr) have been developed. Anadolu University e-Certificate Programs are three-month certificate programs, each of which consists of two or three courses, the education and training are online, and exams are conducted under supervision. Anadolu University e-Certificate Programs started in spring 2007 with an aim to provide continuous education and up-to-date academic content for three semesters a year (Mutlu et al., 2007).

Focusing on personal and professional development Anadolu University e-Certificate Programs offer individuals and businesses a chance to stay one step ahead of their competitors. In addition to that, e-Certificate Programs provide solutions to the issues of businesses that want to hire more qualified staff but cannot find time or opportunity to train their current staff. Along with individuals who are not registered to university, or/and who have completed their university degree but want to improve their knowledge and skills on a different topic, institutions or organizations can also direct their staff who are spread across different regions to Anadolu University e-Certificate Programs and ensure that their employees to pass a standard education. Institutions and organizations can plan their training needs in an efficient, economic, and flexible way by using e-Certificate Programs.

121 different programs in 14 categories are offered on the e-Certificate platform. By far, 122.761 individuals have participated in the e-Certificate Programs and 69.485 have earned certificates. There are no prerequisites to participate in e-Certificate Programs. Anyone interested in subject areas can benefit the programs. Applications can be made online at http://e-sertifika.anadolu.edu.tr Learning materials of e-Certificate programs consist of textbooks prepared according to distance education methods and learning materials on Anadolum e-Campus portal. Specific learning materials, different types of interactive videos and exercises can be reached on the platform, in addition to the unit summary videos, quizzes and pilot tests. Exams of the e-Certificate Programs are conducted online in 81 provinces in Turkey, along with Cyprus and Azerbaijan. With the help of these, the participants can take their exams of the e-Certificate Programs in a city that they stated during their registration application without moving from one place to another.

e-Certificate Programs as career focused lifelong learning platforms provide support to learners in academic, technical, and managerial issues in synchronous and asynchronous environments. Anadolu University e-Certificate Programs that can be reached online offer a wide range of learning opportunities for all ages, education levels and all individuals with any professions ranging from marketing, management, tourism to language training.

Table 2. Distribution of Demographics of the Participants

		n	%
Sex (n=1278)	Female	480	37.6
	Male	798	62.4
Age (n=1243)	18-33	444	35.7
	34-49	616	49.6
	50+	183	14.7
Education Level (n=1278)	Literate	2	0.2
	Primary School	28	2.2
	High School	308	24.1
	Associate Degree	146	11.4
	Bachelor Degree	572	44.8
	Master's Degree	193	15.1
	PhD	29	2.3
Employment Status (n=1279)	Employer	973	76.1
	Not Employer	157	12.3
	Retired	71	5.6
	Student	78	6.1

FINDINGS AND DISCUSSIONS

In this study, "Individuals Preferences in Career Focused Lifelong Learning Communities" questionnaire was sent to 15,318 participants of e-Certificate Programs in the 2019 spring, summer and fall semesters. 1,280 participants responded to the survey and the demographic information of theirs can be seen on Table 2. Many of the students are male with a rate of 62.5% and 37.4% are women. The average age of the students is 44 and the age range is between 34-49 of 51.2%. When the age range is examined, it is seen that there are people between the ages of 18 and 77. This can be considered as evidence that e-Certificate Programs platforms support lifelong learning by providing opportunities to improve themselves. The distribution of the participants based on their education levels showed that 45.2% of the students are on undergraduate level. This can be considered as a sign that well-educated people will re-train themselves when given opportunities. Based on the data of their employment status, 75.9% of the participants were hired. The majority of the students were expected to be hired because of their participation in career focused learning communities.

First of all, the students were asked whether they had previous experience of attendance to such a program/course in order to determine their experiences regarding career focused learning communities (Table 3). Based on the answers, it is seen that the majority of the learners (79.9%) participated in e-Certificate Programs of distance education. This experience is followed with face-to-face education (41.5%) of certificate programs. These certificate programs, which are carried out in a completely face-to-face environment, are generally carried out within the scope of in-house training in institutions where participants work. Then both distance and face-to-face (blended) education programs with certificates (23.7%) followed.

And lastly, the least experiences of participants were (17.7%) related to MOOCs platforms such as Coursera, edX, Udemy. To clarify the difference shortly here; MOOCs and certificate programs are separated from each other as the training is on fee and certification after completing it. The fact that the participant group being more experienced in certificate training can be interpreted as the learners in this group being very eager to participate in lifelong learning activities, as well as they want to certify their training to use in their professional life.

Table 3. Participants' Experiences with Career Focused Learning Communities

		n	%
I have participated in e-certificate programs fully that are provided through distance education. (n=1258)	Yes	1005	79.9
	No	253	20.1
I have participated in certificate programs that are provided through face-to-face education. (n=1255)	Yes	521	41.5
	No	714	56.9
I have participated in certificate programs that are provided by both distance and face-to-face (blended) education. (n=1226)	Yes	291	23.7
	No	935	76.3
I have registered myself to MOOCs platforms like Coursera, edX, Udemy. (n=1239)	Yes	217	17.7
	No	1022	82.5

In the later part, the learners were asked about their motivations to participate in career focused learning communities and offered multi optional questions (Table 4). The responses showed that they participated in these communities mostly to develop themselves (28.8%). However, about their experiences of participating in certificate training, they seemed to prefer certificate training rather than free of charge training since they would like to document their training. One of the top expected motivations to participate was to develop themselves professionally and get promoted at work, but they expressed that actually it was to increase their personal development. While in the literature, "the desire to do their current job in a better way and to get promoted" (Yuan and Powell, 2013, p. 11) is the most stated motivation, it was expressed in third place by the learners participating in this study. These trainings are an option for leisure time (9.1%), but they are in the lower places as a motivation for participation. Career focused learning programs are thought to be preferred by individuals with a reason of them being free or accessible at a small cost for certification (Chen, Barnett and Stephens, 2013; Wilson et al., 2013). Contrary to the literature, the reasonable participation fee was stated as the motivation for participation by a very small group of 3.5%.

Table 4. Motivations of Participants to Participate in Career Focused Learning Communities

	f	%
Increasing my personal development	914	28.8
Increasing my professional development	775	24.4
Progressing in the job that I am currently working in	549	17.3
Certificate being useful to find a job	339	10.7
Make use of my leisure time	290	9.1
Being able to get another job/start another business	200	6.3
Reasonable fee	111	3.5

Table 5. Participants' Communication Media Preferences

	f	%
I learn better in synchronous learning media such as virtual classroom, live course, and face to face counselling.	660	51.8
I prefer a communication medium that fits to my purpose. I use a synchronous medium for chatting, and an asynchronous medium to send my homework.	441	34.2
I prefer email for the messages to be permanent.	325	25.6
I do not use any communication medium. I am only interested in learning materials.	255	19.9
Compared to synchronous media, I express myself better in asynchronous media such as discussion groups and email.	237	18.4

Taking their past experiences into account, the learners were asked which of the synchronous / asynchronous communication media they preferred when participating in a career focused learning program (Table 5). They were offered multichoice questions for their motivations. 24.4% of the students who answered the questionnaire reported a choice of single media, while 75.6% preferred more than one. This result is aligned with the literature, where it is mentioned that no one single technology will be enough to meet the needs of teaching and learning, to satisfy the needs of different learners or to diversify the learning environments. Learners use multimedia synchronously on this platform.

Table 6. Participants' Views on Their Motivations and Expectations

	f	%
I think that the work-focused learning communities I participate in have a positive effect on my motivation for work in general.	899	70.0
I think that the work-focused learning communities I participate in affect my work motivation in a positive way.	915	71.1
I think the work-focused learning communities I attend meet my expectations.	786	61.0

By analysing the distribution of media, it is obvious that the participants with a great rate of 51.8% stated that they learned better in synchronous learning environments such as virtual classroom, live course and face to face counselling. It is thought that the preference of synchronous media is due to the need of learners to communicate by contacting a real person. In her study, which Kip (2007, p. 69) investigated the frequency of open and distance learners using different media support that provided to them, and found out that the students preferred email the most; virtual classroom and live chat the least out of all the academic support services that they were provided such as email, forum and virtual classroom. Whenas, in this study it is found out that asynchronous communication media, such as discussion groups and email, were the least (18.4%) preferred media compared to synchronous ones. The context of the communication is important in this sense. Asynchronous media such as email are preferred for use in task-oriented communications, while synchronous media such as chat in virtual classrooms are mostly preferred for social communication (Hinds and Kiesler, 1995; Walther et al, 2001). One of the findings of this study supports this idea. There are 34.2% of them who chose the option consciously with an approach appropriate for the purpose "I prefer a communication medium that fits my purpose. I use synchronous media for chatting and asynchronous media to send my homework". This was the second most preferred by the learners.

Interestingly, the learners who stated as "I don't use any communication medium. I am only interested in learning materials" constitute a noteworthy ratio of 19.9%. The fact that some of the learners participate in courses without communicating with the institution or other learners can be interpreted as they focus only on the content and take the benefit of this learning system as a course materials platform.

Finally, the participants were asked about their continuity to career focused work communities, and views on their motivations and expectations based on their past experiences (Table 6). 70% of the students stated that the work-focused learning communities they participated in had a positive effect on their work motivation in general. And 71.1% of them stated their job motivations being affected in a positive way; while 61% stated that the work-focused learning communities met their expectations. It is seen that the learners think that the career focused learning communities that they participate in the context of lifelong learning affect their work and job motivation in a positive way and meet their expectations and therefore, they contribute to their personal and professional development.

CONCLUSION

Lifelong learning is a learning process that covers all life in order for the individual to be able to succeed in his social and professional life, to cope with the problems encountered, to provide solutions, to develop competencies he has and to reach the information that is rapidly growing, changing and renewed. Career focused learning communities, in which individuals participate to constantly renew themselves in order to be successful in their social and professional life, provide the opportunity to have lifelong learning experiences through open and distance learning, where they can progress at their own learning pace without sacrificing their daily life. Career focused learning communities have different opportunities in means of synchronous and synchronous communication, in addition to academic, technical and managerial subjects. Individuals communicate with both staff of institutions and other learners by choosing the communication medium that fits their habits and preferences. In this manuscript, it was aimed to determine the communication medium preferences of the individuals participating in a career focused learning community within the scope of lifelong learning. With this general purpose, these were investigated: who participated in these programs, the past experiences of the learners in participating in career focused learning communities, the motivations of participating in these communities, their communication medium preferences and their views on these communities. The preferences of the participants to take advantage of this system that is not compulsory, and they participate to learn are important to investigate to offer suggestions to the future programs about the features that they should provide.

This study was designed in the context of Anadolu University e-Certificate Programs, a career focused lifelong learning platform. These programs are three months online certification programs, each consisting of two or three courses.

Additional support services are provided to learners for academic, technical, and managerial topics in synchronous and asynchronous settings. Anadolu University e-Certificate Programs are the first certification programs in Turkey conducted over the internet; they offer lifelong learning opportunities with a wide range of education catalogue for individuals in all ages, all education levels, reaching out to all professional groups from marketing, management, foreign language to tourism.

In this study, "Individuals Preferences in Career Focused Lifelong Learning Communities" questionnaire was sent to 15,318 participants of e-Certificate Programs in the 2019 spring, summer and fall semesters. 1,280 participants responded to the survey. The demographic information of the participants was given on the first section, their experiences on the second section, their synchronous / asynchronous communication medium preferences when they participate in a career focused learning program on the third section, and learners' views on career focused learning communities on the fourth section. Some of the findings align with the literature while others contradict.

The demographics of the participants indicate that the average age is 44, though the age differs from 18 to 77, and a great majority of them were mostly employed in a job, and were well educated, at the undergraduate level. The age range of the learners can be thought as an indicator that the platforms of e-Certificate Programs provide individuals with the opportunity to improve themselves by ensuring a continuous lifelong learning. It was expected for employed students to consist of the majority as they participated in a career focused learning community. Institutions with certificate learning should provide a wide range of programs in different fields by considering the learner profile and thus support individuals for lifelong learning.

Lifelong learning communities such as e-certificate programs, private courses and MOOCs are of great importance in keeping up to date professional, individual, and social skills in today's world of work. This study showed that the vast majority (79.9%) of the learners have previously participated in e-certificate programs through distance education. This is followed by face-to-face certificate programs, blended certificate programs and the experience of participating MOOCs. To clarify the distinction here; where certification programs and MOOCs differ; all of the certificate training can be summarized as having a fee of registration and learners getting a completion document (certificate) at the end. The fact that the participants of this research are more experienced in certificate programs can be interpreted as that the learners in this participants group are very eager to participate in lifelong learning activities, besides their wish to document their training to use in their professional life. The participation in such classes may be low because of the fact that MOOCs are not that well-known throughout the society. Institutions providing certificate training would ensure that individuals who participate in their courses/programs take advantage of lifelong learning at the highest level if they would guide them to MOOC courses of their own institution's or a different ones.

When why they participated in these platforms is examined, it is seen that the learners participated in such communities in order to increase their personal development the most (28.8%). But this contradicts with their experiences. They stated that they preferred certificate training due to the need for certification rather than free training. While the option to increase professional development and progress in their current job were expected to be the top motivation to prefer participating in these programs, they have stated their top motivation as personal development. A similar contradiction is linked to the registration fee. In the literature, it is stressed that career focused training programs are preferred by individuals because of them being free of charge or accessible at a small cost during the certification stage (Chen, Barnett and Stephens, 2013; Wilson et al., 2013). However, contrary to the literature, the reasonable participation fee was expressed as the least motivation for participation by a very small group of 3.5% in this study. Therefore, it can be said that the participation fee is not a decisive factor for the learners. Similarly, it can be said that the certificate is preferred at low rates in finding a new job or starting another business. This is an indication that individuals consider their certificate training not as a work opportunity but as a lifelong learning experience. From this perspective, institutions providing certificate training should focus on improving personal development opportunities of the learners instead of providing a guarantee to get a job at the end of the training.

When the communication medium preferences of the participants were analysed, it is seen that they learned best in synchronous learning media such as virtual classroom, live courses and face-to-face counselling. The reason to prefer synchronous media is thought due to the needs of learners to communicate with a real person online, and to communicate verbally and visually. Though, asynchronous media, such as email, discussion forums, are found to be the least preferred. But, in career focused learning communities, synchronous media are thought to be preferred more than asynchronous media (Kip, 2007, p. 69). The context of the communication seems to matter here. Asynchronous media such as emails are preferred for task-oriented communications, while synchronous media such as virtual classrooms are mostly preferred for social communication (Hinds and Kiesler, 1995; Walther et al, 2001). A finding of this study agrees with this idea. There is a group of people who chose the option consciously with an approach appropriate for the purpose "I prefer a communication medium that fits my purpose. I use a synchronous medium for chatting, and an asynchronous medium to send my homework.". This option was the second most preferred one by the learners. In the context of the support services offered in career focused learning communities, academic, managerial, and social communications should be examined separately, and the preference of the participants in these contexts while communicating should be investigated by different studies.

On the other hand, there are studies (Fulk et al., 1995; King and Xia, 1997; Walther and Burgoon, 1992; Williams and Wilson, 1999) that one of the most important factors affecting the choice of the medium is the experiences. Similar to that, King and Xia (1997, 168) emphasized that the individual's choice of medium to perform a certain task is closely related to the experience of the person rather than the harmony between the medium and the task. Further studies should be conducted to examine learners' experiences about distance education and the medium that they take place, and to define their communication preferences. It should be noted that experience is decisive in terms of preferences.

It is also seen that there is a considerable group of people that do not use any communication medium but are interested only in learning materials. Some of the learners attend courses without communicating with the institution or other learners, and focus only on the content, they consider this learning system as a course material platform. Considering that the learner having such preferences, virtual libraries can be developed to help those who would be interested in only using the contents on learning platforms. With this way, individuals who prefer to spend time on their own and learn by themselves will meet a variety of learning resources. Here it is also worth to mention that the personality types of the learners will also play a role in the communication medium preferences (Carlson and Zmud, 1999; King and Xia, 1999). While some students express themselves better by speaking, others have a timider character and may prefer to communicate in written media such as email. Taking personality types into account and ensuring the diversity of the environment is important in terms of providing communication media that will embrace all students.

Finally, when the participants' views on career continuity, motivation and expectations based on their past experiences were examined, it was seen that they had a positive effect on their work motivation in general, and they met the expectations of their career focused learning communities they participated in. Under the light of the findings obtained from this manuscript, it will be beneficial to develop communication media based on the preferences of the participants for the career focused learning communities provided by open and distance education.

REFERENCES

Adamopoulos, P. (2013). What makes a great MOOC? An interdisciplinary analysis of student retention in online courses. *Proc. of 34th Intern'l Conf. on Info. Systems*. Retrieved from http://pages.stern.nyu.edu/~padamopo/What%20makes%20a%20great%20MOOC.pdf

Bates, A. (2005). *Technology, e-Learning and Distance Education*. Routledge. doi:10.4324/9780203463772

BFUG. (2003). *Bologna Process between Prague and Berlin, Report to the Ministers of Education of the Signatory Countries*. BFUG.

Breslow, L., Pritchard, D. E., DeBoer, J., Stump, G. S., Ho, A. D., & Seaton, D. T. (2013). Studying learning in the worldwide classroom: Research into edX's first MOOC. *Research and Practice in Assessment*, *8*, 13–25.

Browning, L. D., Satre, A. S., Stephens, K. K., & Sornes, J. O. (2004). *Information & communication technologies in action: Linking theory & narratives of practice*. Copenhagen Business School Press.

Carlson, J. R., & Zmud, R. W. (1999). Channel expansion theory & the experiential nature of media richness perceptions. *Academy of Management Journal*, *42*, 153–170.

CEDEFOP. (2001). *Memorandum on Lifelong Learning Consultation Process: A Review of Member State and EEA Country Reports*. CEDEFOP.

CERI. (2001). Centre for Educational Research and Innovation. Education Policy Analysis. OECD.

Champaign, J., Colvin, K. F., Liu, A., Fredericks, C., Seaton, D., & Pritchard, D. E. (2014). Correlating skill and improvement in 2 MOOCs with a student's time on tasks. *Proc. of First ACM Conf. on Lrng. @ Scale Conf*, 11-20. 10.1145/2556325.2566250

Chen, X., Barnett, D. R., & Stephens, C. (2013). Fad or future: The advantages and challenges of massive open online courses (MOOCs). Presented at Research to Practice Conf. in Adult and Higher Ed, Lindenwood University.

Dave, R. H. (1976). *Foundtions of Lifelong Education*. Pergamon Press.

European Commission. (1994). *White Paper on Growth, Competitiveness, and Employment: The Challenges and Ways Forward into the 21st Century*, Retrieved from https://www.cvce.eu/content/publication/1997/10/13/b0633a76-4cd7-497f-9da1-4db3dbbb56e8/publishable_en.pdf

European Commission. (1995). *European Commission, Teaching and Learning: Towards the Learning Society*. Retrieved from https://europa.eu/documents/comm/white_papers/pdf/com95_590_en.pdf

European Commission. (2002). *European Commission, European Report on Quality of Lifelong Learning, Brussels*. Retrieved from http://www.aic.lv/bolona/Bologna/contrib/EU/report_qual%20LLL.pdf

Fulk, J., Steinfield, C. W., Schmitz, J., & Power, J. G. (1987). A Social Information Processing Model of Media Use in Organizations. *Communication Research, 14*(5), 529–552. doi:10.1177/009365087014005005

Grainger, B. (2013). *Massive Open Online Course (MOOC) Report*. University of London Intern'l. Academy. Retrieved from http://www.londoninternational.ac.uk/sites/default/files/documents/mooc_report-2013.pdf

Güleç, İ., & Çelik, S., & Demirhan, B. (2012). Yaşam Boyu Öğrenme Nedir? Kavram ve Kapsami Üzerine Bir Değerlendirme. *Sakarya University Journal of Education, 2*(3), 34–48.

Gündoğan, N. (2003). Avrupa Birliği'ne Üye Ülkelerde Bir İstihdam Politikası Aracı Olarak "Yaşamboyu Öğrenme" ve Bazı Örnek Program ve Uygulamalar. *Kamu İş, 7*(2).

Heinich, R., Molenda, M., Russell, J., & Smaldino, S. (1999). *Instructional Media and Technologies for Learning* (6th ed.). Prentice-Hall.

Hinds, P., & Kiesler, S. (1995). Communication across boundaries: Work, structure, & use of communication technologies in a large organization. *Organization Science, 6*(4), 373–393. doi:10.1287/orsc.6.4.373

Holmes, A. (2002). *Lifelong learning*. Capstone Publishing.

Jarvis, P. (2004). *Adult Education and Lifelong Learning: Theory and Practice* (3rd ed.). Routledge. doi:10.4324/9780203561560

King, R. C., & Xia, W. (1999). Media appropriateness: Effects of experience on communication media choice. In K. E. Kendall (Ed.), *Emerging information technologies: Improving decision, cooperation, & infrastructure* (pp. 143–175). Sage. doi:10.4135/9781483345505.n7

Kip, B. (2007). *Çevrimiçi Öğenenlerin Farklı Destek Ortamlarını Kullanma Sıklıklarıyla Sosyal Bulunuşluk Algıları Arasındaki İlişki* (Unpublished master's thesis). Anadolu University, Eskişehir.

Kulich, J. (1982). Lifelong education and the universities: A Canadian perspective. *International Journal of Lifelong Education, 1*(2), 123–142. doi:10.1080/0260137820010204

Moore, M., & Kearsley, G. (2005). *Distance education: a systems view*. Wadsworth.

Mutlu, M. E., Özöğüt Erorta, Ö., Kip Kayabaş, B., & Kayabaş, İ. (2014). Anadolu Üniversitesi Açıköğretim Sisteminde e-Öğrenmenin Gelişimi. In Açıköğretimle 30 Yıl (pp. 1-58). Anadolu University Publications.

Newman, J., & Oh, S. (2014, June 13). 8 things you should know about MOOCs. *Chron. of Higher Ed.* Retrieved from http://chroniclc.com/article/8-Things-You-Should-Know- About/146901/

OECD. (2004). *Lifelong Learning, Policy Brief.* Retrieved from http://www.oecd.org/ dataoecd/17/11/29478789.pdf

Reiser, R., & Gagne, R. (1983). Selecting Media for Instruction. Academic Press.

Seaton, D. T., Bergner, Y., Chuang, I., Mitros, P., & Pritchard, D. E. (2014). Who does what in a massive open online course? *Communications of the ACM*, *57*(4), 58–65. doi:10.1145/2500876

Tan, C. L., & Morris, J. S. (2006). Undergraduate college students, laptop computers, and lifelong learning. *The Journal of General Education*, *54*(4), 316–338. doi:10.1353/jge.2006.0013

Toprak, M., & Erdoğan, A. (2012). Yaşamboyu öğrenme: Kavram, politika, araçlar ve uygulama. *Yükseköğretim ve Bilim*, *2*(2), 69–91.

Vail, R. (2013). *Who is afraid of a big bad MOOC?* Colorado Mesa University. Retrieved from http://cmubiznews.org/uploads/facultyinsights/Vail_Who_is_afraid_of_a_MOOC_Febru ary_2013.pdf

Wain, K. (1993). Lifelong education: Illiberal and repressive? *Educational Philosophy and Theory*, *26*(1), 58–70. doi:10.1111/j.1469-5812.1993.tb00186.x

Walter, J. B., & Burgoon, J. K. (1992). Relational Communication in Computer-Mediated Interaction. *Human Communication Research*, *19*(1), 50–88. doi:10.1111/j.1468-2958.1992.tb00295.x

Wang, Y., & Baker, R. (2015). Content or platform: Why do students complete MOOCs? *Journal of Online Learning and Teaching*, *11*(1), 17.

White, J. P. (1982). *The aims of education re-stated.* Routledge & Kegan Paul.

Williams, S. R., & Wilson, R. L. (1999). Group support systems, power & influence in an organisation: A field study. In K. E. Kendall (Ed.), *Emerging information technologies: Improving decisions, cooperation & infrastructure*. Sage. doi:10.4135/9781483345505.n8

Wilson, B., Orr, J., Hansen, D., Khoja, S., Bryant, R., & Iyer, G. (2013). MOOCs and their impact on CS education. *Journal of Comp. Sci. in Colleges*, *29*(1), 110–112.

World Bank. (2003). *Lifelong Learning in the Global Knowledge Economy: Challenges for Developing Countries*. Retrieved from https://documents.worldbank.org/curated/en/528131468749957131/Lifelong-learning-in-the-global-knowledge-economy-challenges-for-developing-countries

Yuan, L., & Powell, S. (2013). MOOCs and open education: Implications for higher education. Bolton, UK: JISC, Centre for Educational Technology & Interoperability Standards.

KEY TERMS AND DEFINITIONS

E-Certificate Programs: Short-term education programs for personal and professional development of individuals designed on the basis of lifelong learning philosophy.

Lifelong Learning: A learning process that continues throughout one's whole life in order to reach the rapidly developing, changing and renewed information that will help the individual to succeed in social and professional life, cope with the problems, provide solutions when needed, and develop his competencies.

Massive Open Online Courses: Courses that are generally free, designed for large groups of students, accessible by anyone with an internet connection, and without any entry requirements.

Open and Distance Learning: Learning process in which learners are distant from each other and learning resources in terms of time and/or space, and their interactions with each other and learning resources are based on remote communication systems.

Chapter 10
Research and Network Opportunities in the Field of Science Communication

Rabia Tas
Anadolu University, Turkey

ABSTRACT

Science and technology have been developing at a great pace. There are numerous breakthrough innovations and groundbreaking improvements in almost every aspect of life. Undoubtedly, another field experiencing such a fast development is communication. Communication has been following and adapting related technological developments not only in terms of software and hardware but also in terms of targeting the right audience and creating the audience interaction. While developments in science and technology feed communication, communication disseminates the outcomes of innovative studies to the defined target groups. Science communication becomes a part of this mutual relation. This study is focused on research funds, opportunities, and international networks in the field of science communication. Within this context, challenges in research activities and expected impact of research outcomes will be examined in terms of science communication. Also, the role and importance of networks in science communication will be examined in line with research funds and support provided for this field.

DOI: 10.4018/978-1-7998-4534-8.ch010

INTRODUCTION

Science and technology have been developing at a great pace. History of science and technology is consisted of breakthrough innovations and groundbreaking improvements in almost every aspect of life. Undoubtedly, another field which has been experiencing such a fast development process is communication. Communication has been following and adapting related technological developments and inventions not only in terms of software and hardware but also in terms of targeting the right audience and creating the audience interaction. Increasing number of media such as social media tools and digital environments bring new approaches to language, visuals, content, and such. Therefore, it's inevitable to bring science and technology together with the communication. While developments in science and technology feed the developments in communication, communication disseminates the outcomes of innovative studies to the defined target groups. At this point, science communication becomes a part of this mutual relation. Science communication is the way of transferring knowledge of science-related subject matters to the society by using right tools and approaches. There are a lot of research funds and organizations supporting scientific studies and technological initiatives. Communication as a field of research can benefit from these opportunities. However; science communication as a field of study has gained more and more importance as the number of inventions and the necessity of innovation arise. Therefore; specific research calls and funds have become available for science communication. In addition, networks and non-governmental organizations to support and promote science communication have been established for many reasons including creating awareness and drawing attention to the importance of the subject. This study is focused on existing research funds and opportunities in the field of science communication as well as the international networks in this field. Within this context, the challenges that need to be addressed in research activities and the expected impact of the outcomes of these activities will be examined in terms of science communication. Also, the role and importance of networks in science communication will be examined in line with these research funds and support provided specifically for this field.

Science Communication

There are various definitions for science communication. Davies and Horst (2016) defined science communication as "organized actions aiming to communicate scientific knowledge, methodology, processes or practices in settings where non-scientists are a recognized part of the audience" (p. 4). Mc Kinnon and Vos (2016) described it as "encompassing the interactions between the various publics and the scientific community at a popular level, which may include some form of educational or pedagogic approach" (p. 29). Polino and Castelfranchi (2012) made an assumption about key factors of science communication based on their literature review and described it as "to explain (to share) scientific knowledge or ideas with nonspecialized audiences" (p. 7). These definitions and other publications in literature enables people to make different definitions and explanations on the concept. In general science communication can be stated as using appropriate communication tools as a way of telling complex scientific knowledge in a simple narration.

Some people may show a general interest in science communication to follow recent developments and news on the latest science and technology activities while some people consider it as a field of study in itself. Those considering science communication in a professional way can benefit from existing funds and network opportunities in this field. There are various types and forms of support mechanisms for science communication all around the world.

Research Opportunities in the Field of Science Communication

Science and society relations are among the areas on which researchers focus and develop theories as well as practices. Communicating developments in science and technology is one of those research areas. The general name of these studies can be defined as science communication, but it is possible to see different wordings such as public communication or it is possible to see names in relation to their focus area such as STEM communication.

As in other research and developments fields, studies on science communications are also funded through grants and research opportunities provided by different organizations. Here are some of the research opportunities on science communication around the world which was randomly found on search engines by making query with keywords *science communication research* and *science communication grants*:

Horizon 2020 - Science with and for Society

European Commission has been funding the program named Horizon 2020 since 2014. Horizon 2020 has been announced as the biggest European Union Research and Innovation programme so far. The program budget for over 7 years is almost €80 billion. The main goal of the program is to support breakthrough innovations, discoveries and world-firsts by taking ideas from laboratory stages to market level (European Commission, n.d.). Horizon 2020 program has different work programs under different fields of research such as energy, security and transport. One of these fields is *Science with and for Society*.

Science with and for Society is addressing the societal challenges of Europe. Its goals are to make science more attractive for society and especially to young population, to attract society's attention to innovation, and to provide support for further research and innovation activities. This program approach is called Responsible Research and Innovation (RRI) and it is open for all related actors including researchers, citizens, policy makers, business people, etc. So far, three work programs have been published for Science with and for Society with numerous research projects calls within. These programs are categorized in terms of the years in action: 2014-2015 Work Programme, 2016-2017 Work Programme and 2018-2020 Work Programme (Horizon 2020, n.d.). Different project calls have been announced under these work programs and some projects have been funded.

In the Work Programme 2014-2015, there was a support call for The Euroscience Open Forum (ESOF) 2016 whose expected impact was to form a conference that would be one of the biggest European science communication platforms (Horizon 2020 Work Programme 2014 – 2015, n.d., p. 47). The Euroscience Open Forum (ESOF) was founded in 2004 by EuroScience and it is the biennial pan-European meeting specified on research and innovation. At ESOF meetings scientists, researchers, young researchers, business people, entrepreneurs and innovators, policy makers, science and technology communicators and the general public from all over Europe get together to discuss developments in the sciences, humanities and social sciences (The Euroscience Open Forum, n.d.). The Science with and for Society Program of Horizon 2020 has become a funding mechanism for this meeting to take place.

In the Work Programme 2016-2017 there was a support call for The Euroscience Open Forum (ESOF) 2018 whose expected impact was once again to form a conference that would be one of the biggest European science communication platforms (Horizon 2020 Work Programme 2016 - 2017, n.d., p. 62). The Euroscience Open Forum (ESOF) supported by this program was held in Manchester in 2016 and in Toulouse in 2018 (ESOF, n.d.).

In the Work Programme 2018-2020 there was a support call numbered SwafS-19-2018-2019-2020 and named Taking stock and re-examining the role of science communication. The scope of this call was defined as "to better understand how results from research and scientific methodologies are communicated and perceived by citizens (taking into account age, gender, and socio-economic status), to develop improved ways to measure and assess science communication, and to identify good practices and policy guidelines to increase the accuracy of (and therefore trust in) science communication." Its expected impact was increase quality and efficiency of communication and interaction between scientists, media and public in general (Horizon 2020 Work Programme 2018 - 2020, n.d., pp. 56-57). Also there was a support call for The Euroscience Open Forum (ESOF) 2020 which will be held in Trieste in September 2020 (Horizon 2020 Work Programme 2018 - 2020, n.d., pp. 70-71). In this work program, European Commission provided a specific call for science communication research apart from The Euroscience Open Forum

(ESOF) support calls. This can be seen as an indicator of developments in science communication in the society. Calling for innovative projects that address this challenge can be evaluated as a need in society to fill the interaction gap between scientists and general public to achieve a more engaged knowledge society in Europe.

European Research Council Awards

European Research Council supports investigator-driven frontier research. Researchers from all fields can apply for European Research Council funds. The funds provided by this Council not only support individual research projects but also contribute to create a strong European research system (European Research Council, n.d.). The Council has announced a call for an award named The European Research Council's (ERC) Public Engagement with Research Award 2020. This award was planned to recognize and appreciate the efforts of previous European Research Council grantees for their efforts in public engagement and outreach. The engagement with public under this award call means communication, consultation and collaboration with public. The categories were (1) public outreach, (2) press and media relations and (3) online and social media (ERC Award, n.d.). This award is not for research and innovation funding, but for recognizing the successful efforts and emphasizing the importance of science communication in society.

European Research Council also has its own Project Promotion Team and it has a *#Science Communication* page where webcomics, visual features and other forms of communication have been used to explain previously funded projects and their outcomes to the public (ERC Science Communication, n.d.). This page is an example of social engagement by practicing it at an organizational level. Individual researchers and other scientists can take these applications as good practices to inspire them for further communication efforts.

Austrian Science Fund Programs

The Austrian Science Fund (FWF) is Austria's main organization providing funds for basic research (FWF Der Wissenschaftsfonds, n.d.). This organization has been offering calls for researchers who previously received a grant from The Austrian Science Fund (FWF) to take measures for science communication in relation to the research project funded by the FWF. The budget for this 12 month grant can be up to 50.000 € per application (FWF – WissKomm, n.d.). This is an example of a national fund for increasing science communication and creating awareness in this field as well. This type of grants for science communication activities can also be embedded in the original project calls announced by research and innovation councils all around the world.

Swiss National Science Foundation Programs

The Swiss National Science Foundation (SNSF) is a private foundation established in 1952 (SNSF, n.d.). This foundation supports science communication activities between researchers and between researchers and the publis in general. It has different grant and support programs for science communication such as media training, dialogue between researchers and Agora- taking research to the public (SNSF Science Communication, n.d.). Agora program aims to support dialogue between scientists and society by bringing researchers' current studies to an audience of lay people. The program enables a platform where scientists and people can get together and communicate through interacting and listening to each other (SNSF Agora, n.d.). This is an example of a private foundation to support science communication in a country as a complementary alternatives for national research programs and grants. This model can be replicable to different countries in different methods or approaches.

Natural Sciences and Engineering Research Council of Canada

Natural Sciences and Engineering Research Council of Canada has a pilot program to fund Science Communication Skills Training for 1 year period, up to a $20.000 budget. They define science communication activities critical for Canadian science culture. Therefore; the Council offers this award program for students, fellows, and faculty in science, technology, engineering and math (STEM) from Canadian postsecondary institutions (NSERC Canada, n.d.). This grant is provided for receiving a training on science communication. Therefore; it will have a long-term effect on the grantees for their career in science and technology. They will have a chance to gain a skill to base their efforts in engaging with society.

TUBITAK Science Youth

The Scientific and Technological Research Council of Turkey (TUBITAK) was established in 1963 as the leading agency for management, funding and conduct of research in Turkey. Its main goal is to advance scientific and technological researches of Turkishh scientists (TUBITAK, n.d.). TUBITAK Bilim Genc (Science Youth) is an initiative to support science communication in Turkey through news, interviews and many other activities aiming at transferring scientific knowledge to public. This initiative doesn't offer a budget for science communication research. However; it helps to disseminate the research and its outcomes through different tools. Science Talks in Turkey, news about projects made in schools and universities, visual documentaries of research projects are among the services that this initiative offers (TUBITAK Bilim Genc, n.d.). This practice is a model for engaging younger generations to talk about their scientific effort and to communicate through a national institution's support and encouragement. This also provides a recognition to researchers and their contribution to Turkish science and technology ecosystem.

These programs from Europe, Austria, Sweden, Canada and Turkey are just examples of some research funds and supports around the world. These examples can be increased in number and type. The importance of these supports are to give a chance for researchers to further their studies and to keep their motivations high. Science communication is a scientific field itself and academic studies will feed the improvement of this field.

Network Opportunities in the Field of Science Communication

Networking has become one of the most effective tools of today's global environments. There are different types of networks bringing people all around the world together to focus on a common interest. Business networks, research networks, education networks, mentor networks and many other networks can be listed among these initiatives. The definition of *network* is originally related with information communications technology (ICT) and defined as "a system of lines or channels resembling a network" such as telephone network or computer network (Network, Merriam-Webster, n.d.). In time, bringing different channels and systems together has been defined as networks apart from ICT fields as well. The term social network has begun to be used more frequently. Networks' ability to connect various parts of a system together is accepted as a capability which can be used in other contexts. Therefore; *networking* has become a part of social environments and this noun has been defined as "the activity of meeting people who might be useful to know, especially in your job" by putting it into a vocational context (Networking, Cambridge Dictionary, n.d.). In recent years, there has been a lot of get-togethers named networking events, networking meetings, networking happy hours, etc. It has also become a profession for some people and the term *networker* has been used to address those professionals.

Why networking has become such an important part of professional lives? It is possible to answer this question with multiple answers. However; the basic answer could be framed around globalization and joint interests effecting people all around the world. Since we're living in an interrelated world, it's almost impossible to ignore developments in other countries performed by other researchers and developers. Therefore; we need to be aware of the recent actions in subject matters that we are interested in; and if there is a way for us to contribute to those actions, we need to be able to get in touch with the related parties. It is a mutual effort and other people all around may need to contact us to cooperate and collaborate. Even though communication technologies have developed to make these connections possible, it is hard to follow every development in each and every country. People focusing on a specific interest have their own acquaintances all around the world, but it is a limited and personal data to keep. In order to maximize this data, people need clusters on a specific subject to work, discuss and develop together. That is one the main reasons why networking has become an important part of professional lives.

Corporates, non-governmental organizations (NGOs), national and international organizations, governments and many other actors can start an initiative of a network. For instance; European Commission has initiated The European Cooperation in Science and Technology (COST) and it has been receiving European Union funds under different programs since 1971. The main goal of COST is to fund organizations to create research networks under specific COST actions to support research and innovation activities in a specific field for 4 years. Thanks to this research network; researchers from universities, laboratories, private companies and other related institutions can represent their countries in a joint research endeavor to overcome a specific problem such as diagnosis and treatment of some diseases or empowering the social initiatives and such (COST, n.d.). The types and goals of networks can change and the many other examples can be added to the list of networks.

There are networks in the field of science communication bringing international parties together around this subject matter. Some of them are regional organizations' initiative while some of them are national or institutional efforts. Here are some of the networks on science communication around the world which was randomly found on search engines by making query with keywords *science communication network*:

Science Communication Network

Science Communication Network was established in 2002 as a non-profit organization and its target group consists of journalists, decision makers and public in general. Their goal is to create awareness in environmental and public health issues. This network is based in the United States of America and it received "Champion of Environmental Health Research Award" from the National Institute of Environmental Health Sciences (NIEHS) for their contribution to bringing scientific knowledge to public through journalists. Their activities are mainly focused on interacting scientists and journalists to bring attraction to scientific facts by using different interaction tools such as interviews and trainings (Science Communication Network, n.d.). It is known that terminology in the field of health sciences is hard to be understood by general public. Therefore; doctors and pharmacists simplify their explanations to convey their messages related with health and treatment. Since journalists aren't generally coming from this background, they need to convey news about environment and public health in a more comprehensible way compared to the terminological words of experts. From this point of view, Science Communication Network has been working as a bridge between these two different worlds to create a meaningful outcome to be served to the general public.

The efforts of conveying environment and public health related messages and news in accordance with the profile of the audience can affect different people in different ways. Students can get fascinated by some news they read on newspaper or an interview they watch on TV. This interaction can affect their decisions to choose a career profession. People waiting for a treatment to a specific illness can read a doctor's explanation on recent developments in that treatment and this knowledge can provide moral support to the healing process. Academic papers and scientific publications can feed the need to share current knowledge in research and innovation. However; their target audience is mostly limited with other researchers working on similar subject matters. But these kinds of papers need to be translated into a different terminology and wording to be understood by a general audience. By doing this, it prevents spread of fake news and misinformation.

Network for the Public Communication of Science and Technology

Network for the Public Communication of Science and Technology is an association whose objectives are mainly to improve theoretical studies on science communication and create a platform for practitioners and theoreticians to study and perform science communication practices (PCST Network, n.d., p. 3). This network is based in Australia and they have a legal status as an incorporated association Members of this network come from different backgrounds including researchers, science centers and museum staff, journalists, students, writers, web designers, visual and performing artists and professionals working on science themes. It has a Scientific Committee and they conduct a bi-annual conference (Network for the Public Communication of Science and Technology, n.d.). Their conference series started in 1989. The conferences include presentations, reflections on practice, workshops and demonstrations (PCST Conferences, n.d.). This international network in public communication of science and technology has been contributing to science communication in an academic and research based approach while providing reflections on practices.

Science communication is a multidisciplinary field. It can be addressed by social science researchers as well as physical science, life sciences and formal sciences. Therefore; it is important to provide tools and platforms for them to come together to work on this subject from different point of views and approaches. Communication between members of the association and the interaction among the participants of the conferences contributes to the literature of science communication. Sharing good practices and recent developments can maximize the impact and they can help to disseminate the outcomes of studies and researches. By doing this Network for the

Public Communication of Science and Technology offers a platform to nourish and nurture the scientific studies on science communication.

One of the conferences of Network for the Public Communication of Science and Technology was held in Istanbul, Turkey on 26-28 April 2016. The conference's host institute in Turkey was Hacettepe University. During this event, a web page publishing only in Turkish was designed and set up. The address of that web site is www.bilimiletisimi.com (Bilgici, 2016). The web site is still available and it has been posting news about scientific developments and information from national and international sources such as newspapers, organizational web sites and social media posts. This outcome of the conference is an example and an indicator of its sustainability and capacity of dissemination. This kind of practices and their impact can multiply with the increase of the numbers in such networking platforms. Also digitalization of science communication practices is another benefit for reaching out those sites and posts through different mediums.

European Science Communication Institute

European Science Communication Institute (ESCI) is a non-profit organization empowering scientists in terms of communication to bring science and society closer. The Institute was established in 2016 and it supports national and international research initiatives by maximizing their dissemination efforts as well as providing trainings and other support tools for scientists (European Science Communication Institute, n.d.). The Institute is an established organization rather than a social network; but its efforts and contributions connect researchers and other related parties under a framework. Therefore; it can be considered among the network opportunities in the field of science communication.

The Institute's services are given under eight categories which are visual identity, science journalism, TV broadcast, social media, web videos, citizen engagement, behavior change and communication training. These services have different target groups and methods of practice; but their main goal is to be a bridge between science and society (ESCI, n.d.). It is possible for scientists to contact the Institute to disseminate their projects and to find a service that addresses their needs. This type of institutional supports can be replicable in other countries or regions. In that perspective, they can be defined as role models to institutions to create their own outlet for transferring scientific knowledge into public information.

Science Communication Work Group

Science Communication Work Group is a platform for EuroScience members to develop their communication skills, to exchange of ideas and practices and to provide a field for dialogue about science communication. EuroScience is a European based, non-profit grassroots association. It has a wide target group which consists of researchers, teachers, students, science administrators, policy-makers and general public that shows interest to science, technology and innovation. The main goal is to link science and technology to the society. Also they are open to all ages, disciplines and nationalities as well as different sectors like public and private sector. This work group under EuroScience addresses scientists that are good at research but have challenges to express their research outcomes to the general public. Therefore; they provide good practices, tips and workshops to overcome this challenge. They have social media accounts to reach out to their audience and the they share the archive of previous tip sheets (Science Communication Work Group, n.d.). This is a practical platform for people from different backgrounds to get a gist of general challenges in science communications and to find some tips to use for finding solutions to the communication problems.

Science Communication Work Group doesn't describe themselves as a *network;* but they can be listed among them since they link people together on a specific subject matter. Their social media accounts have followers and these people can learn from each other through their experiences and best practices. Also EuroScience brings this work group together at workshops in during European Open Forum (ESOF). It is possible to download tip sheets prepared to offer practical solutions to science communication challenges. Also these tip sheets include the contact information of the researcher preparing these tips. Therefore; it is a good opportunity for networking and furthering the interaction with the researcher whose studies you are interested in.

QTurkey Community

QTurkey Community is a network focusing specifically on quantum related events and activities in Turkey. Its target group is students all around the country. They organize events like "Introduction to Quantum Programming" workshops. However; one of their goals is to create popular science content to Turkish readers interested in quantum related activities. They define themselves as a communication and collaboration platform that brings quantum-enthusiasts together (QTurkey, n.d.). This community is an example of a network with limited target group and a limited subject matter. This type of local or national networks can ignite an interest in a specific field to be communicated with the society and in time, it can become a part of a bigger network which can be either a national or an international initiative. They designed their science communication efforts under a platform named "Quantum Technologies Popular Content Pool". The goal of this pool is to create quantum related content which is scientifically accurate, free of copyright, written in Turkish, peer-reviewed and ready to be used in any science communication and popular science platform (QTurkey Bilim İletişimi, n.d.).

This community addresses students which are interested in quantum related developments. These students will become professionals after graduation. Therefore; joining a science communication network during school years can help them acquire a culture to transfer their scientific knowledge into a public information which addresses a general public profile rather than the professionals in their field. This learning experience by becoming a part of a network is a practice-based learning for them and a valuable opportunity. Taking such approaches as good practices, students can be encouraged to establish student clubs or student networks to communicate their theoretical knowledge into practical popular science contents for general audience.

These networks from the United States of America, Australia, Europe and Turkey are just examples of science communication networks around the world. These examples can be increased in number and type. The importance of these network opportunities is to fill a gap between scientists who are very successful in their professions but not as successful in their communication skills and people in general audience who are interested in scientific and technological developments but not experts in vocational terminologies. The knowledge is everywhere. Communication is an inseparable part of our daily lives. Using those basic sources and combining them with different tools on different subject matters can increase the level of interaction in the society.

CONCLUSION

Science and technology are among the most important parts of today's world. With the effect of globalization, the studies in these areas are above borders and above disciplines. Developments and improvements require public engagement at all levels. Life sciences require cooperation and collaboration with social sciences. The world is an interwoven network of knowledge and information. In this situation, it's crucial to communicate with people for people. User experiences, feedbacks, expectations and needs should be communicated between the scientists and the public in general. The way to accomplish this interaction is to develop skills in science communication.

Research funds and networks in science communication are very valuable sources for those who are interested in this field. The increase in science communication practices will provide people valuable input and opportunity to observe the effects of it. Science and society meeting will further scientific activities by enabling the society to understand and share their opinions about those scientific studies. Also it will speed up the adoption of technological and scientific outcomes in daily lives by breaking down the prejudices. Communication is key for every successful initiatives which include science and technology as well.

REFERENCES

Agora, S. N. S. F. (n.d.). *Agora - where research meets the public.* Retrieved from http://www.snf.ch/en/funding/science-communication/agora/Pages/default.aspx

Award, E. R. C. (n.d.). *Public Engagement With Research Award.* Retrieved from https://erc.europa.eu/managing-your-project/public-engagement-research-award

Bilgici, Z. (2016). *Bilim İletişimi.* Retrieved from https://bilimgenc.tubitak.gov.tr/makale/bilim-iletisimi

Conferences, P. C. S. T. (n.d.). *History of PCST conferences.* Retrieved from https://www.pcst.co/conferences

COST. (n.d.). *About COST.* Retrieved from https://www.cost.eu/who-we-are/about-cost/

Davies, S. R., & Horst, M. (2016). *Science Communication: Culture, Identity and Citizenship*. Palgrave Macmillan. doi:10.1057/978-1-137-50366-4

ERC Science Communication. (n.d.). *#Science Communication.* Retrieved from https://erc.europa.eu/tags/science-communication

ESCI. (n.d.). *European Science Communication Institute – Our Services.* Retrieved from https://www.esci.eu/service/

ESOF. (n.d.). *Past ESOF.* Retrieved from https://www.esof.eu/en/about/esof-in-trieste.html

European Commission. (n.d.). *Horizon 2020.* Retrieved from https://ec.europa.eu/programmes/horizon2020/what-horizon-2020

European Research Council. (n.d.). *Mission.* Retrieved from https://erc.europa.eu/about-erc/mission

European Science Communication Institute. (n.d.). *European Science Communication Institute - About Us.* Retrieved from https://www.esci.eu/about-us/

FWF WissKomm. (n.d.). *Science Communication Programme (WissKomm).* Retrieved from https://www.fwf.ac.at/en/research-funding/fwf-programmes/science-communication-programme-wisskomm/

Horizon 2020. (n.d.). *Science with and for Society.* Retrieved from https://ec.europa.eu/programmes/horizon2020/en/h2020-section/science-and-society

Horizon 2020 Work Programme 2014 - 2015. (n.d.). *16. Science with and for Society Revised.* Retrieved from https://ec.europa.eu/research/participants/data/ref/h2020/wp/2014_2015/main/h2020-wp1415-swfs_en.pdf

Horizon 2020 Work Programme 2016 - 2017. (n.d.). *16. Science with and for Society.* Retrieved from https://ec.europa.eu/research/participants/data/ref/h2020/wp/2016_2017/main/h2020-wp1617-swfs_en.pdf

Horizon 2020 Work Programme 2018 - 2020. (n.d.). *16. Science with and for Society.* Retrieved from https://ec.europa.eu/research/participants/data/ref/h2020/wp/2018-2020/main/h2020-wp1820-swfs_en.pdf

McKinnon, M., & Vos, J. (2016). Crossing A Threshold. In M. C. A. van der Sanden & M. J. de Vries (Eds.), Science and Technology Education and Communication: Seeking Energy (pp. 27-45). Sense Publishers.

Network for the Public Communication of Science and Technology. (n.d.). *About the PCST Network.* Retrieved from https://www.pcst.co/about

Network. (n.d.). In *Merriam-Webster dictionary online*. Retrieved from https://www.merriam-webster.com/dictionary/network

Networking. (n.d.). In *Cambridge Dictionary online*. Retrieved from https://dictionary.cambridge.org/dictionary/english/networking

NSERC Canada. (n.d.). *Science Communication Skills grant (pilot).* Retrieved from https://www.nserc-crsng.gc.ca/Promoter-Promotion/ScienceComm_eng.asp

PCST Network. (n.d.). *The PCST Network Constitution.* Retrieved from https://www.pcst.co/doc/PCST.Constitution.2016-04-27.pdf

Polino, C., & Castelfranchi, Y. (2012). The 'Communicative Turn' in Contemporary Techno-science: Latin American Approaches and Global Tendencies. In B. Schiele, M. Claessens, & S. Shi (Eds.), *Science Communication in the World: Practices, Theories and Trends* (pp. 3–17). Springer. doi:10.1007/978-94-007-4279-6_1

QTurkey. (n.d.). *About QTurkey Community.* Retrieved from https://www.qturkey.org/about

QTurkey Bilim İletişimi. (n.d.). *Science Communication.* Retrieved from https://www.qturkey.org/science-communication

Science Communication Network. (n.d.). *Science Communication Network: Who We Are, What We Do*. Retrieved from http://sciencecommunicationnetwork.org/

Science Communication Work Group. (n.d.). *Science Communication Work Group Presentation*. Retrieved from https://www.euroscience.org/structure/science-communication-workgroup/

SNSF. (n.d.). *Mission*. Retrieved from http://www.snf.ch/en/theSNSF/organisation/Pages/default.aspx

SNSF Science Communication. (n.d.). *Science Communication*. Retrieved from http://www.snf.ch/en/funding/science-communication/Pages/default.aspx

The Euroscience Open Forum. (n.d.). *ESOF in a Nutshell*. Retrieved from https://www.esof.eu/en/about-esof/aims.html

TUBITAK. (n.d.). *Who We Are?* Retrieved from https://www.tubitak.gov.tr/en/about-us/content-who-we-are

TUBITAK Bilim Genc. (n.d.). *Bilim Genc*. Retrieved from https://bilimgenc.tubitak.gov.tr/dergimiz/hakkimizda-0

Chapter 11
Scientific Research Articles:
Twenty-Two Language Errors to Avoid

John Blake
https://orcid.org/0000-0002-3150-4995
University of Aizu, Japan

ABSTRACT

Error-free scientific research articles are more likely to be accepted for publication than those permeated with errors. This chapter identifies, describes, and explains how to avoid 22 common language errors. Scientists need to master the genre of scientific writing to conform to the generic expectations of the community of practice. Based on a systematic analysis of the pedagogic literature, five categories of errors were identified in scientific research articles namely accuracy, brevity, clarity, objectivity, and formality. To gain a more in-depth understanding of the errors, a corpus investigation of scientific articles was conducted. A corpus of 200 draft research articles submitted for internal review at a research institute with university status was compiled, annotated, and analyzed. This investigation showed empirically the types of errors within these categories that may impinge on publication success. In total, 22 specific types of language errors were identified. These errors are explained, and ways for overcoming each of them are described.

INTRODUCTION

English is the *de facto* language of scientific communication (Swales, 1997; Simionescu and Simion, 2004). This means that scientists, regardless of origin and mother tongue, who aspire to publish world-class research need to read and write scientific research articles in English (Ventola, 1994; Bitetti and Ferreras, 2017).

DOI: 10.4018/978-1-7998-4534-8.ch011

Publishing in top-tier English language scientific journals is a prerequisite for career advancement in many countries, or as Ventola (1992) puts it, "publish in English or perish" (p.191). Graduation from doctoral programs, gaining a university post, drawing down research funding and securing tenure may all be contingent on having been published in journals of a pre-determined rank or impact factor (Schein, Farndon and Fingerhut, 2000; Peat, Elliott, Baur and Keena, 2002).

Despite the focus on novelty and substance of research contributions, research articles that are permeated with lexical, grammatical or genre-related errors are more likely to be rejected. Intrusive errors may cause reviewers to misunderstand the intended meaning, severely lowering the chance for acceptance of a manuscript submitted for publication. One non-intrusive language error is unlikely to result in rejection, but multiple non-intrusive errors may create a negative impression possibly leading the academic gatekeepers to fall victim to the horns-and-halo effect, and assume that the research itself may also contain errors (Maiorana and Mayer, 2018). In short, writing error-free prose reduces the possibility of rejection.

This chapter focuses on the language difficulties that scientists need to overcome to write scientific research articles that conform to the generic expectations of the community of practice (Lave and Wenger, 1991). By pinpointing the potential points of failure, i.e. the language errors, scientists can take proactive measures to avoid making similar errors, and adopt a systematic approach to identify these types of errors in their own writing.

Twenty-two specific types of language errors are identified, explained and suggestions for addressing them are given. As there are no native speakers of scientific communication, these errors apply to both native and non-native English speakers. Although this chapter limits its claims to English and scientific writing, it is likely that similar types of errors are present in formal scientific texts written in other languages, and in other formal writing genres, such as academic and research writing.

This chapter starts by reviewing the literature on language errors with particular reference to scientific writing and scientific research articles. However, among the various classification systems that are discussed, no system focusses on the language errors that potentially lead to rejection by reviewers. The following section describes the attempt to address this lacuna in the research by conducting a template analysis of the pedagogic literature. The next section describes the annotation and analysis of a corpus of draft research articles collected at a research institute in Japan. Starting with the five categories identified by the template analysis, the corpus investigation revealed a total of 22 types of errors. The following sections describe, explain and suggest ways to address these errors. The chapter concludes by summarizing the key concepts and identifying future research directions.

The primary objectives of this chapter are to:

- show the importance of error-free writing to scientists and researchers
- demonstrate how genre, practitioner expectations and scientific writing are intertwined
- detail a study to investigate the type of errors in scientific writing
- describe, explain and exemplify the type of errors discovered
- provide a taxonomy of language-related errors
- exemplify and explain each type of error
- suggest ways to avoid or ameliorate these errors

BACKGROUND

Scientific writing is characterized by its lexical complexity, lexical density (Halliday and Martin, 1993; Hyland, 2006) and informational density (Holtz, 2009). The use of compact syntactic structure and the extensive use of nominalization (Halliday & Martin, 1993; Halliday, 1994; Holtz, 2009; Biber and Gray, 2013) result in texts that are dense and often impenetrable to lay readers (Hyland, 2006; Klein *et al.*, 2017). Despite the linguistic and technical complexity of this genre, errors that occur in scientific writing are essentially the same types of errors that occur in academic writing. Common advice from journal editors is to use professional editing services or ask a native English speaker to proofread. Researchers in non-Anglophone countries may not have access to either the funds for professional service or native-speaking colleagues (Canagarajah, 1996). There is also considerable variation in the relative difficulty of learning English with mother tongue being one of the key factors. Languages that are more linguistically distant (i.e. less similar) are thought to be more difficult to acquire (Richards and Schmidt, 2002). Therefore, it may be particularly difficult for scientists whose first language is radically different to English, such as Chinese and Japanese, to write error-free scientific texts.

Errors in writing can be categorized in numerous ways, such as by the source of the error (Fries, 1945; Lado, 1957; Selinker, 1972; Brown, 2000; Wee, 2009), the structure or form of the error (James, 1998; Nesselhauf, 2003), the relative or absolute frequency (Orr and Yamazaki, 2004), the language proficiency of the learner (Selinker, 1972; Grefen, 1979; Edge, 1990) or semantic impact (Burt and Kiparsky,1974; Elliot, 1983).

Source error can be attributed to either a person (e.g. teacher) or language (e.g. mother tongue) that is considered to be the source of the error. The source of the error may stem from the learners themselves (learner-induced) or be the result of the teaching received (teacher-induced) (Ferris, 2006). Intralingual errors (Fries, 1945, Lado, 1957; Selinker, 1972; Brown, 2000) are those that originate in the target language itself while interlingual errors are errors that are ascribed to interference from another language, which in many cases is the mother tongue. Intralingual errors include *inter alia* simplification, overgeneralization, hypercorrection and fossilization (Touchie, 1986).

Edge (1990) extended the works of earlier researchers, such as Gefen (1979) by proposing a classification system that focuses on the relationship between learner proficiency and error type. The slip-error-attempt system advocated by Edge (*ibid.*) can be used by teachers to assess the type of error correction that could be used. Accidental slips are errors that learners can self correct, ingrained errors are those that learners or their peers at the same proficiency level may be aware of while attempts are errors that are beyond the learners' proficiency level, which means that teacher correction is recommended. Ferris (2011) suggests the dichotomy of treatable (rule-based errors) versus untreatable (idiosyncratic) errors.

Errors may also be categorized by the form of written text to which they may be determined to affect. Three common candidate error types are lexical, grammatical and discoursal. Nesselhauf (2003) delineates errors on a scale from morphemic to lexical to syntactic. Lexical errors can themselves be further classified into three categories namely mis-selection, misformation and distortion (James, 1998, p.145). Mis-selection refers to the wrong word choice, misinformation refers to words that exist in the first language but not the target language while distortion refers to words that do not exist in either the first or target languages. Some linguists, particularly systemic functional linguists, may argue that lexical and grammatical should be combined into one category namely lexicogrammatical since grammar itself is realized through words. Orr & Yamazaki (2004) identified twenty errors made in graduation theses drafted by computer science majors at a public university in Japan. The majority of these errors were grammatical located within noun phrases (e.g. articles) or verb phrases (e.g. tense and voice).

Errors that stop the reader understanding the intended meaning or confuse the reader are intrusive (Elliot, 1983) while errors that do not affect the meaning and readers simply ignore are non-intrusive. Although presented as a dichotomy, the reality is a cline from non-intrusive to intrusive. Others researchers (e.g. Burt and Kiparsky,1974; Bates, Lane and Lange,1993) use the term *local* for non-intrusive errors and *global* for intrusive errors.

To assist writers aiming to secure publication, errors that may lead to the rejection of a research article are the focus of attention. These errors are the same types of errors that are likely to result in failing grades should the manuscript be submitted to university faculty. Given the importance of the genre of research communications, many researchers have investigated the reasons that reviewers and editors of journals and conference proceedings reject manuscripts. The four reasons commonly reported as the main causes for rejection are that the research is unoriginal, unimportant, methodologically flawed or plagued by poor language (Bordage, 2001; Pierson, 2004;

McKercher *et al.* 2007; Thrower, 2012; van Weijen, 2014). The first three errors are primarily the concern of the subject specialist while the fourth error, poor language, is the topic of this chapter. Research articles in scientific domains are unlikely to be rejected because of one or two minor language errors. However, intrusive errors may confuse reviewers and stop them understanding the article. Numerous non-intrusive errors may cause negative associations. In short, both intrusive and non-intrusive may reduce the chance that a research article is accepted for publication.

The focus of this chapter is on errors that are more likely to lead to a submission being rejected by reviewers.To identify these critical errors, it is necessary to understand the types or errors that tend to occur in scientific research articles. This knowledge can be used to create a taxonomy that can be used to create checklist for scientists to proofread their manuscripts or serve as the basis for the development of a more complete framework to support writers of scientific research articles. The study was divided into two phases: template analysis and corpus investigation

The starting point for phase one was a survey of the pedagogic literature, which was followed by a template analysis. A detailed description of the procedure and interim results is provided in the template analysis section. The results of the template analysis were then used as the starting point for the subsequent corpus investigation.

PHASE ONE: TEMPLATE ANALYSIS

This investigation was conducted *in situ* at a research institute located on the north-coast of *Honshu*, the largest of the Japanese islands. The postgraduate institution has university status and offers research degrees at masters and doctoral levels. Research is conducted in three disciplines: materials science, information science and knowledge science. The institute is well resourced with library materials in both English and Japanese, and a dedicated writing resource center. The official website claims that the university is bilingual, but interviews with students and faculty revealed that the reality was more bipart-lingual with the Japanese-speaking and the English-speaking communities having limited interaction.

A thorough survey of the pedagogic materials available to support the scientific writing of researchers in this research institute was conducted. Each reference book in the writing section was skimmed to discover whether the contents included information on scientific or research writing. Books that contained at least some reference to scientific research writing were added to the list of resources to be analyzed.

In total, fifteen English language books on scientific communication in the resource center met the criteria for selection for inclusion in the analysis. Each book was read by the author and relevant concepts were coded using template analysis (King, 2004). Template analysis begins with a set of predetermined codes (e.g. grammar, spelling, etc.), but permits *ad hoc* amendment to the initial codes and addition of supplementary codes throughout the study. Template analysis can be viewed as midway between grounded theory with no *a priori* codes and content analysis which prescribes all codes *a priori*.

At the conclusion of the template analysis, the final tagset comprised five categories of errors. The three major categories were accuracy, brevity and clarity; the remaining two minor categories were (perceived) objectivity and formality. Table 1 lists the resource books that were included in the analysis and shows which of the categories in the final five-category tagset were mentioned in each book. Once the template analysis had been completed for the published materials, the highest-ranked websites identified by the Google algorithm for relevant keywords suggested by a focus group of researchers (e.g. scientific writing) were also checked. to see whether their results aligned with those discovered during template analysis.

Although these five categories were identified from the pedagogic literature *per se*, this does not necessarily mean that these categories are reflected in published or draft scientific research articles. The disjuncture between materials designed to teacher English for specific or academic purposes and authentic materials is well documented (e.g. Blake, 2015). In the corpus investigation described in the following section, the applicability of these pedagogic categories is tested and further refined.

PHASE TWO: CORPUS INVESTIGATION

Following the template analysis, a corpus investigation of draft scientific articles was conducted to investigate empirically the types of errors that impinge on publication success. This corpus investigation utilised the findings from the template analysis. The first aim is to ascertain the validity of the results of the template analysis. The garbage-in garbage-out maxim holds. If the content of the pedagogic literature does not reflect reality, then there will be some divergence between the corpus results and the results of the template analysis. The second aim is to extend the granularity of the categories so that teachers of scientific writing may more effectively support novice writers. Frontline teachers of scientific writing who are familiar with the specific types of common errors that writers make can more easily design syllabuses, materials and lessons to help their students avoiding making the same types of errors.

A corpus of two-hundred draft research articles submitted for internal review at the same research institute was compiled. Most authors were PhD candidates, but a small proportion were either faculty or master degree students. Mother tongues of the first authors represented in the corpus include Japanese, Thai, Chinese and Vietnamese.

Table 1. Ethnographic study of the writing resource books published in English in a large research institute in Japan.

Author Surname	Year	Title (abbreviated)	Accuracy	Brevity	Clarity	Objectivity	Formality
Alley	1996	The craft of scientific writing	A	B	C		
Bailey	2014	Academic writing	A		C		
Bentley	2003	Report writing in business	A				
Browner	2012	Publishing and presenting clinical research		B	C		
Fearing & Sparrow	1989	Technical writing: Theory and practice			C	O	
Flowerdew	2014	Academic discourse	A	B	C	O	F
Graves & Graves	2012	A strategic guide to technical communication			C		
Hall	2011	How to write a paper	A	B	C		
Hartley & Buckmann	2001	Business communication	A	B	C		
Laan	2012	The insider's guide to technical writing	A		C		
Ober	2007	Contemporary business communication	A	B	C		
O'Connor	2002	Writing successfully in Science	A		C		
Markel	2012	Technical communication			C		
Matthews & Matthews	2007	Successful scientific writing	A	B	C		
Sageev & Romanowski	2001	A message from recent engineering graduates in the workplace					F
		Totals	10	7	13	2	2

Errors in the corpus were identified manually and semi-automatically using rule-based parsing. On discovering an error, the error was annotated with a tag from the developing tagset. The tagset evolved over the annotation with error types being added and combined until the tagset stabilized. An in-depth description of the methodology of this investigation is available in Blake (2018).

The five pedagogic categories were found to be suitable and functioned well to classify the types of errors in the draft research article corpus. Each of the five categories was further divided into between three and five subcategories, hereafter referred to as types of error. In total, twenty-two types of errors were found in the corpus. The most frequent errors identified were overwhelmingly related to factual or grammatical accuracy. The types of errors are listed below.

Six types of accuracy errors were identified, namely:

1. factual errors in world knowledge,
2. factual errors within the article,

3. overly bold claims,
4. overgeneralization errors,
5. spelling and grammar errors, and
6. statistical and numerical errors.

Errors relating to brevity included:

7. using multiple vague words,
8. repeated words, and
9. redundant words.

There were five types of clarity errors, specifically:

10. use of vague expressions,
11. lexical ambiguity,
12. referential ambiguity,
13. syntactic ambiguity, and
14. garden path sentences.

Errors related to perceived objectivity were categorized into three types:

15. focus on people and feelings rather than things and ideas,
16. emotive wording, and
17. excessive personalization.

The final category of formality errors consists of five types of error, namely:

18. apostrophes,
19. abbreviations,
20. slang,
21. informal terms, and
22. rhetorical questions.

In the following sections, each of these errors is discussed in detail. Concrete examples of the errors have been extracted from the corpus of draft research articles, and where necessary anonymized by substituting specialist terminology with XXX. The extracts contextualize the error description, the error type itself is then described and explained. Practical suggestions are provided to help life-long learners avoid making similar types of errors.

The following five sections detail the twenty-two error types using the error categories of accuracy, brevity, clarity, objectivity and formality. The Pareto principle held for the corpus results with the accuracy category comprising approximately 80% of all the errors identified.

ACCURACY ERRORS

Some of the errors in this category are likely to be considered the most serious by reviewers. The core aim of academic and scientific writing is to convey data, information and/or knowledge with zero distortion of meaning, and so errors that cause degradation of meaning need to be avoided.

Error type 1: Factual errors in world knowledge
Corpus example: The population of Japan is 12,734,100 [1].

There are two main citation styles: author-year and numeric (as in this example). The corpus example is a citation from the source listed as the first item in the reference list indicated by the number in square brackets. There is, however, an error within the citation itself. When copying the number in figures, the writer omitted the final zero, and so the population became 10 times smaller. This type of careless error can cause severe confusion. In this example, readers may either believe the incorrect figure or work out that the figure is incorrect. Through this careless copying error, the writer may have created a negative impression, which could lead the reviewer to succumb to the cognitive bias known as the horn effect and be unduly influenced by this error (MacDougall, Riley, Cameron and McKinstry, 2008). Copying errors are easy to avoid by being systematic. For example, when using the copy-and-paste function, doublecheck that the first and last words or numbers have been included. It is more difficult to make selection errors when using the arrow keys to control the cursor than manipulating a mouse, which relies on fine motor skills, and so the arrow keys are the recommended method of selection. Should it be necessary to copy onto paper, make sure that not only the final numbers or words are included, but that other copying errors do not creep in due to ambiguity caused by poor handwriting. Sixes and zeros may be mistaken for each other, particularly when zeros are not exact ovals.

Error type 2: Factual errors within the article
Corpus example: There are two types of.... First, ... Second, ... Third, ...

When writing a research article, the content of the draft may change significantly. This could be due to changes decided by the writer, but also may be the result of changes suggested or insisted on by co-authors, reviewers, editors or proofreaders. Mistakes in lists are commonplace. The number of items in the list should reflect the number of items that follow. For example, in the list above, initially there were only two items, but apparently a third item was added after the second without changing the introductory sentence which still only lists two items. This type of error appears obvious, but is easily missed by writers. However, people who read this without having seen earlier versions can easily notice such errors.

Error type 3: Overly bold claims
Corpus example: XXX will play a key factor in the near future.

The past has already occurred, the present is occurring now, but the future has yet to come. Although we may attempt to predict the future, we cannot be certain. Even simple events in the future may not happen due to unforeseen circumstances. The modal verb *will* tends to be used to show the certainty of a future action or state. In the example above, the writer expresses certainty. This could become true, but should a natural disaster, war or other event happen, would the claim still hold? Claims that are overly bold can be hedged in three ways. First, the certainty level can be reduced to show a degree of probability, e.g. using modal verbs such as: *could*, *may* or *might*. Second, the scope of the claim can be limited to one that will be certain (or at least no-one could prove otherwise). Third, a condition can be attached to the claim, to add a proviso.

Error type 4: Overgeneralization errors
Corpus example: All women…

Claims about a characteristic of all members of a particular group are problematic. Is it possible to finish the corpus example in a way that is not contentious? Here are some suggestions to consider.

a. All women are not men.
b. All women have two X chromosomes.
c. All women are human.

The first claim, a, presents an either-or dichotomy in which women cannot be men. In logic the law of noncontradiction states that two propositions *A is B* and *A is not B* cannot both be true. However, this law assumes shared definitions of both *A* and *B*. Definitions of women and men vary depending on religious, political or scientific beliefs. Thus, the term *women* can be interpreted in different ways depending on the beliefs related to biological sex and gender, it can be argued that a person born *a* biological male can undertake gender reassignment and become a woman. The second claim, *b*, is also problematic as some people with Swyer syndrome are born with XY chromosomes but have female genitalia and identify as female. The third claim, *c*, is not contentious, since by definition the term women has three elements of meaning: *female, adult, human* and *plural*. When making claims about a groups, it is necessary to play devil's advocate and identify whether it is possible to argue against the claim. If it is, then consider adding modality, limiting the scope or adding a conditional.

Error type 5: Spelling and grammar errors, especially for users of LaTeX
Corpus example: There are three form the first experiment that XXX.

Users of WYSIWYG text processing software, such as Microsoft Word, are used to receiving help from spelling and grammar checkers. Problems can, however, occur when the language settings of the document or sections of the document are not set to English. For example, when pasting text from a non-English document, it is possible to accidently change the language settings, which means that the English words in that section will not be checked. The situation appears even worse for users of LaTeX (Lamport, 1994), which is a plain text editor. This means that writers do not see what their text looks like until it is compiled into a pdf. A key problem for novice uses of LaTeX is not using a spellchecker, since with many free LaTeX editors, spellcheck is not included by default. In the corpus example, Microsoft Word was unable to identify that *form* should have been *from*. Checking spelling and grammar in one's own writing is particularly difficult given the screen-memory interference issue. In short, writers may not necessarily read what is currently written but may read what had been written.

Error type 6: Statistical and numerical errors
Corpus example: ($p < 0.5$)

The probability value, p, is used to show the probability of finding equivalent or more extreme results while assuming the null hypothesis is true. In short, the smaller the p value the higher the statistical significance. However, reported p values tend to be either $p < 0.05$ or $p < 0.01$. These values are often considered sufficient to claim statistical significance when conducting hypothesis testing. In this example, the mistake was a typographic error that was not noticed in proofreading. The probability of 50% means that the result is random and no significance can be claimed, and the article will most likely be rejected. This careless oversight is similar to the errors discovered in medical records. Hanauer *et al.* (2019) found numerous numerical errors including 1972 invalid dates such as January 35 and February 30. There were also 128 instances of patients aged 150 years old.

BREVITY ERRORS

Error type 7: Using multiple vague words
Corpus example: The concept that was chosen as the focus of this research is XXX.

Genres may be placed on a cline from terse to verbose. Terse genres, such as scientific research articles are concise, but may be considered difficult to read. Verbose genres may be easier to read and understand, but use more words. To quote Faber (2017), "no one wants to read excessively long studies". Words that do not add substance to the meaning should be omitted. Abstract nouns, such as *concept*, are rather vague. In this corpus example, *concept* refers cataphorically to *XXX*, and as such adds no meaning *per se,* and thus the first six words can be deleted.

Error type 8: Repeated words
Corpus example: We analyze XXX regarding the XXX qualities, XXX qualities and XXX qualities.

Introductory sentences for sections within a research article, tend to list the contents of the section. The corpus example introduces a section that describes three types of qualities. However, the word *qualities* is unnecessarily repeated three times. This sentence can be improved simply by using the word *qualities* once at the head of a noun phrase and then listing the three items as modifiers in the tail of the prepositional phrase that modifies the head. Following this suggestion, the revised concise sentence would be:

We analyze XXX regarding the qualities of XXX, XXX and XXX.

Error type 9: Redundant words
Corpus example: On each and every occurrence, the XXX was noted.

The determiners *each* and *every* have similar but not identical meanings. However, either one on its own is sufficient. They may be used together to show emphasis or to solve translation issues, particularly in legal documents. Legal doublets (Ingels, 2006) are created when near synonyms linked together with a conjunction, which is usually *and*. Cases in point include *aid and abet*, *terms and conditions, goods and chattels*, and so on. Although it can be argued, that only one term would suffice, selecting just one near-synonym may result in the target language translation differing in meaning from the source language as the words are near and not exact synonyms. However, in scientific writing, it is unlikely that a researcher needs to stipulate both each occurrence and every occurrence, and so in this case either term would suffice.

CLARITY ERRORS

Error type 10: Use of vague expressions
Corpus example: This is something which is XXX from XXX.

The pronoun *something* is vague as it can refer to so many different items. Clarity in written English is determined by the specificity of the terminology selected. On a scale from clear to obscure, indefinite pronouns, such as *something*, *someone* or *somewhere*, limit their reference to a type of item, namely thing, person or place; but are rather obscure. The clarity of *something* can be increased by substituting a more specific class of item, such as *feature*, *aspect*, *artifact*, *component*, or so forth. Items can be visualized on a cline from vague to clear, as in this example, which shows the cline of obscurity-clarity from *thing* to *toy poodle*.

thing > animal > mammal > dog > poodle > toy poodle

Error type 11: Lexical ambiguity
Corpus example: It is really high for XXX.

In this example *really* is not vague but it has two possible specific meanings. Either it shows the meaning "to a high degree" and can be replaced by the adverb *very* or is used to state the truth or fact of the situation and so could be replaced by the adverb *actually*. However, it is not possible to understand from this decontextualized example (or in fact the example within the original draft research article) what the intended meaning was. This ambiguity can be resolved by replacing *really* with either *actually* or *very* to reflect the intended meaning. In addition, *really* is rather informal, and so fails to meet the genre expectations for the degree of formality.

Error type 12: Referential ambiguity
Corpus example: Referring to Smith [10], Jones notes that he…

The ambiguity is introduced by the third person pronoun *he* which refers anaphorically to an earlier-mentioned person. However, in this context, there are two candidate antecedents, which creates confusion. Referential ambiguity also commonly occurs when pronouns, especially *it* and *this,* are used anaphorically. This type of ambiguity can be resolved by avoiding the need for a pronoun. Thus, the example sentence could be written without the use of a pronoun, thereby alleviating the need to ascertain the antecedent of the pronoun.

a. *Jones notes that Smith [10] ...*
b. *Smith [10] notes that Jones ...*

Error type 13: Syntactic ambiguity
Corpus example: XXX found two AAA and one BBB, which CCC.

The relative clause *which CCC* may modify only one item namely *BBB* or could modify three items namely *two AAA and one BBB*. This ambiguity is caused by using a modifier after noun phrases linked by the conjunction *and*. Resolution is straightforward. One way to limit the modifier to one noun phrase is to bring the noun phrase to the start of a series and place the modifier immediately after that phrase. The other way is when the modifier applies to each item in a series, and so adding a distributive determiner, such as *all* removes any ambiguity. The corpus example could be rewritten as one of the following depending on the intended meaning.

a. *XXX found one BBB, which CCC, and two AAA.*
b. *XXX found two AAA and one BBB, all of which CCC.*

Error type 14: Garden-path sentences
Author-created example: The journal plans to publish your paper were just a rumour.

It was not possible to anonymize the garden-path sentences in the corpus, and so this example sentence was created specifically to exemplify the issue. On reading this sentence, most readers will assume that *plans* is a verb, but on reaching the finite verb *were*, re-read the sentence and re-categorize plans as a *noun*. Garden path sentences are not only ambiguous, but the initially assumed meaning is not the intended meaning. Comedians use this ambiguity for humour (Dynel, 2009), but reviewers find little humour in being confused. Garden path sentences are difficult for writers to notice because the writer of the sentence is primed to focus on the intended meaning. Second readers or proofreaders can, however, easily notice garden path sentences because they need to backtrack to figure out the intended meaning. The difficulty with garden path sentences is not their revision and disambiguation, but their initial identification.

OBJECTIVITY ERRORS

Error type 15: Focus on people and feelings rather than things and ideas
Corpus example: I will describe the results of our research in section 2.

The corpus example uses the first person pronoun *I* and the possessive adjective *our*. This directs attention to the people involved in the research rather than the research itself. This is in contrast to the expectations of the community of practice that focus on things and ideas rather than people and feelings. Research in the many fields within the humanities and social sciences has embraced the use of first person pronouns, *we* and *I,* and no longer eschews active voice in favour of passive voice. However, research in many applied and pure science domains continues to prefer research articles that are appear "objective" in terms of language usage. The corpus example can be depersonalized and written using just five words as follows:

Section 2 describes the results

Error type 16: Emotive wording
Corpus example: We are pleased to announce that the results show XXX.

In the same vein as the previous example, this corpus example does not adhere to disciplinary expectations of "objectivity". The use of the first-person plural pronoun *we* in some scientific domains may be acceptable, but the adjective *pleased* expresses happiness and, as such, focuses on feelings rather than ideas. The author might have been attempting to frame the result announcement in a manner so as to direct the reader to view the results positively. The sentence can be revised by omitted the emotive sentence stem as shown below:

Results show XXX.

Error type 17: Excessive personalization
Corpus example: ...such as services to your XXX, to your XXX, and to XXX.

Disciplinary expectations vary among pure and applied sciences, but excessive personalization, such as the overuse of personal pronouns or possessive adjectives is likely to convey the impressive that the research article is not "objective" enough. This "objectivity" refers to the appearance of objectivity through depersonalization of the research narrative. Hyland (2002b) notes that teachers of writing to students with English as an additional language, tend to direct them to depersonalize texts by removing references to themselves from their texts. There is a continuum of usage of first-person *I* in scientific and academic writing. Tang and John (1992) identified six functions realized by the use of first-person *I*, namely: representative, guide, architect, recounter, opinion holder and originator. The opinion holder function seems at odds to the desire to present factual information objectively. However, Hyland (2002a) found that expert writers used I when promoting their own work which was in direct contrast to novice writers who used I to, for example, describe the organization of their research article. The strongest authorial presences created using first person pronouns are when elaborating arguments and stating results or claims (Hyland, 2002a). Perhaps, well-established authors attempt to appeal to their own authority to persuade readers of the validity of their work. However, this is a technique that few would advise novice writers to attempt.

The corpus example can be revised as shown in *a*. This in turn can be made more concise by removing the repeated word, *to*. The final version is shown in *b*.

a. …such as services to XXX, to XXX, and to XXX.
b. *…such as services to XXX, XXX, and XXX.*

FORMALITY ERRORS

Error type 18: Apostrophes
Corpus example: To be more precise, XXX doesn't directly cause the effect (E).

The first type of formality error involves apostrophes. There are two cases when the usage of apostrophes is rather informal. First, when they are used to mark the omission of letters within a word. Second, when apostrophes are used in conjunction with the letter *s* to indicate possession. Revisions are straightforward in both cases. Rather than using contractions, the full form should be used and so *doesn't* should be written as *does not.* Kingdom (2014, p.55) states that "the best approach is to avoid them whenever possible". When apostrophes are used to show possession, the noun phrases can usually be reversed and the conjunction *of* inserted between. For example, *the experiment's rate* becomes *the rate of the experiment.* Another apostrophe problem that affects users of scripts that use different character sets is the type of apostrophe. In the Japanese character set, the apostrophe is close to the western apostrophe, but differs in shape (e.g. ` cf. ').

Error type 19: Abbreviations
Corpus example: This is the RQ of this paper.

The introduction section of many research articles introduces the research question, hypothesis, purpose, aims and/or objective. Students and teachers of research methodology may refer to research questions as RQ, but this abbreviation is neither standard nor common. If the author introduces the abbreviation in brackets after the full form, then the abbreviation can be used later in the research paper. However, unknown or undefined abbreviations should be avoided. Research articles are written from experts in a particular field who share some basic background domain knowledge. This basic knowledge may involve the use of abbreviations. These abbreviations may be acceptable, but it is necessary to check the specific disciplinary discourse as there is significant variation. If there is no advice or precedents to the contrary, the safest option is to avoid using unknown or undefined abbreviations. It should be noted that there is no need to put abbreviations after terms that will not be referred to again later in the research article.

Error type 20: Slang
Corpus example: A bunch of IT engineers collaborated and launched…

Slang describes the type of language used only in specific social circles. These social circles may be related a range of demographic factors, such as age and social status. Slang may creep into scientific writing either through lack of awareness of the linguistic register or poor proofreading. Non-native English speakers who acquire language in Anglophone settings by focussing on improving their spoken communication skills, such as the author of the corpus example; may not realize the inappropriacy of their lexical choices, which in this case is the noun *bunch*. Slang and colloquial language was found by Hanauer et al. (2019) in official medical records. Examples of slang found included *hell of a lot*, *waist deep in* and *boat loads*. Slang errors are easily revised by substituting a neutral or formal term for the offending word. A suggested revision of the corpus example is given below.

Corpus example: A team of IT engineers collaborated and launched…

Error type 21: Informal terms
Corpus example: They launched the website right after the earthquake…

Informal terms differ from slang in that their usage is more widespread and not limited to particular social groups. A common example would be using the word *yeah* instead of *yes*. In the corpus example, the informal adverb *right* is used to mean *immediately*. Informal terms invariably can be replaced with more formal equivalent terms in the same way as slang errors are resolved. The resolution of the corpus example is given in *a.* To avoid any potential referential ambiguity with the plural pronoun *they*, passive voice could be used to rewrite the sentence as in *b.*

a. They launched the website immediately after the earthquake…
b. The website was launched immediately after the earthquake…

Error type 22: Rhetorical questions
Corpus example: The key question to ask is: how can we XXX ?

Rhetorical questions are taught in many high school English classes as a way of interacting with readers. Magazine articles and blogs frequently make use of such devices. Rhetorical questions, however, are less common in scientific research articles, although they are used in editorials. Novice writers would be advised to avoid such techniques, but this is not to say that they are never used. Well established professors do make use of rhetorical questions to lead readers through arguments by encouraging readers to participate in thought experiments and consider options. Rhetorical questions are not commonly used as hooks to attract interest. Rather than posing a question, the author should state the answer to the question. Thus, the corpus example could be reworded as:

XXX can be achieved by XXXX.

FUTURE RESEARCH DIRECTIONS

This research identified a taxonomy of errors that is intended to help scientists draft research articles that are free of language-related errors. The taxonomy provides a theoretical foundation on which resources can be created to support scientific writers.

Of the extant software designed for scientific writing, SWAN, the Scientific Writing AssistaNt is particularly useful (Kinnunen *et al.* 2012). SWAN is a downloadable Java software package that uses a rule-based system to provide feedback to writers using quality metrics. However, there is a niche for an online tool that specifically focuses on providing genre-specific feedback to writers on the language used in scientific texts.

This typology of twenty-two errors served as the theoretical foundation for the creation of a prototype error detection tool (Blake, 2018; Blake, in press) for written scientific communications. The future research direction of this research is the practical application of the taxonomy to help scientific writers. This approach is twofold. First is the refinement of the automatic identification of errors to achieve increased precision using both rule-based parsing and probabilistic parsing. Second is the development of a constructive feedback system harnessing multimodal resources that enables writers to correct the errors identified in an efficient and effective manner.

CONCLUSION

The key outcome of this study was the identification of the list of twenty-two common error types categorized into five broad categories. The categories of accuracy, brevity, clarity, objectivity and formality can serve as filters through which written work can be checked. The twenty-two error type taxonomy can be used to serve as a checklist which can be used to proofread scientific writing for language-related errors. This list is complemented with numerous examples of errors extracted from the corpus of draft research articles.

These error categories were created purely for pedagogic purposes. At times, some of the category boundaries are rather fuzzy, meaning that some errors may be caught by more than one category. For some research purposes, this would be a negative outcome. However, the aim of proofreading is to catch all errors, and so providing more than one opportunity to discover an error, is not necessarily a negative outcome.

The taxonomy also acts as a framework for the development of an automatic error detection tool. This software architecture of the tool was created to find the twenty-two error types. Software developers write functions to address each of the errors. At present, the state-of-the-art research in computational linguistics is not able to address all the types of errors, but progress is being and will continue to be made.

Life-long learners armed with knowledge of common errors made in scientific research articles are better placed to be able to reflect on and identify errors in their own writing.

ACKNOWLEDGMENT

This research received no specific grant from any funding agency in the public, commercial, or not-for-profit sectors.

REFERENCES

Alley, M. (1996). *The craft of scientific writing*. Springer. doi:10.1007/978-1-4757-2482-0

Bailey, S. (2014). *Academic writing: A handbook for international students*. Routledge. doi:10.4324/9781315768960

Bates, E., Lane, L., & Lange, J. (1993). *Writing clearly: Responding to ESL compositions*. Heinle & Heinle Publishers.

Bentley, T. (2003). *Report Writing in Business*. Elsevier.

Biber, D., & Gray, B. (2013). Nominalizing the verb phrase in scientific writing. In B. Aarts, J. Close, G. Leech, & S. Wallis (Eds.), *The verb phrase in English: Investigating recent language change with corpora* (pp. 99–132). Cambridge University. doi:10.1017/CBO9781139060998.006

Blake, J. (2018). Corpus-based error detector for Computer Science. In *Proceedings of the Fourth Asia Pacific Corpus Linguistics Conference* (pp.50-54). Takamatsu.

Blake, J. (2020). Genre-specific error detection with multimodal feedback. *RELC Journal, 51*(1), 179-187. https://doi.org/10.1177/0033688219898282

Blake, J. (2015). Prescriptive-descriptive disjuncture: Rhetorical organisation of research abstracts in information science. In F. Formato & A. Hardie (Eds.), *Conference proceedings of 8th International Corpus linguistics Conference* (pp. 377–8). Lancaster: Lancaster University.

Bordage, G. (2001). Reasons reviewers reject and accept manuscripts: The strengths and weaknesses in medical education reports. *Academic Medicine*, *76*(9), 889–896. doi:10.1097/00001888-200109000-00010 PMID:11553504

Brown, H. (2000). *Principles of Language Learning and Teaching*. Prentice Hall.

Browner, W. S. (2012). *Publishing and Presenting Clinical Research* (3rd ed.). Lippincott Williams & Wilkins.

Burt, M., & Kiparsky, C. (1978). Global and local mistakes. In J. Schumann & N. Stenson (Eds.), *New Frontiers in Second Language Learning* (pp. 71–79). Newbury House.

Canagarajah, A. S. (1996). "Nondiscursive" requirements in academic publishing, material resources of periphery scholars, and the politics of knowledge production. *Written Communication*, *13*(4), 435–472. doi:10.1177/0741088396013004001

Di Bitetti, M. S., & Ferreras, J. A. (2017). Publish (in English) or perish: The effect on citation rate of using languages other than English in scientific publications. *Ambio*, *46*(1), 121–127. doi:10.100713280-016-0820-7 PMID:27686730

Duly, H., & Burt, M. (1974). Natural sequences in child second language acquisition. *Language Learning*, *24*, 23–40.

Dynel, M. (2009). *Humorous garden-paths: A pragmatic-cognitive study*. Cambridge Scholars Publishing.

Edge, J. (1990). *Mistakes and correction*. Longman.

Elliott, A. B. (1983). *Errors in English*. National University of Singapore Press.

Faber, J. (2017). Writing scientific manuscripts: Most common mistakes. *Dental Press Journal of Orthodontics*, *22*(5), 113–117. doi:10.1590/2177-6709.22.5.113-117.sar PMID:29160351

Fearing, B. E., & Sparrow, W. K. (Eds.). (1989). *Technical writing: Theory and practice*. Modern Language Association.

Ferris, D. R. (2006). Does error feedback help student writers? New evidence on the short-and long-term effects of written error correction. In K. Hyland & F. Hyland (Eds.), *Feedback in second language writing: Contexts and issues* (pp. 81–104)., doi:10.1017/CBO9781139524742.007

Ferris, D. R. (2011). *Treatment of error in second language student writing*. The University of Michigan Press. doi:10.3998/mpub.2173290

Flowerdew, J. (2014). *Academic Discourse*. Routledge. doi:10.4324/9781315838069

Fries, C. (1945). *Teaching and Learning English as a Foreign Language*. University of Michigan.

Gefen, R. (1979). The analysis of pupil's errors. *English Teachers'. Journal*, *22*, 16–24.

Graves, H., & Graves, R. (2012). *A Strategic Guide to Technical Communication*. Broadview.

Hall, G. M. (2011). *How to Write a Paper* (5th ed.). Wiley – Blackwell.

Halliday, M. A. K. (1994). *An Introduction to Functional Grammar* (2nd ed.). Edward Arnold.

Halliday, M. A. K., & Martin, J. R. (1993). *Writing science: Literacy and discursive power*. The Falmer Press.

Hanauer, D. A., Mei, Q., Vydiswaran, V. V., Singh, K., Landis-Lewis, Z., & Weng, C. (2019). Complexities, variations, and errors of numbering within clinical notes: The potential impact on information extraction and cohort-identification. *BMC Medical Informatics and Decision Making*, *19*(3), 75. doi:10.118612911-019-0784-1 PMID:30944012

Holtz, M. (2009). Nominalization in scientific discourse: A corpus-based study of abstracts and research articles. In *Proceedings of the 5th Corpus Linguistics Conference, #341*. Available at: http://ucrel.lancs.ac.uk/publications/cl2009/

Hyland, K. (2002a). Authority and invisibility: Authorial identity in academic writing. *Journal of Pragmatics*, *34*(8), 1091–1112. doi:10.1016/S0378-2166(02)00035-8

Hyland, K. (2002b). Options of identity in academic writing. *ELT Journal*, *56*(4), 351–358. doi:10.1093/elt/56.4.351

Hyland, K. (2006). *English for Academic Purposes: An Advanced Resource Book*. Routledge. doi:10.4324/9780203006603

Ingels, M. B. (2006). *Legal English communication skills*. Acco.

James, C. (1998). *Errors in language learning and use*. Longman.

King, N. (2004). Using templates in the thematic analysis of text. In C. Cassell & G. Symon (Eds.), *Essential guide to qualitative methods in organizational research* (pp. 256–270). Sage. doi:10.4135/9781446280119.n21

Kingdom, W. (2014). English Grammar and Style: Good writing Practice. *Medical Writing*, *23*(1), 55–57. doi:10.1179/2047480613Z.000000000172

Kinnunen, T., Leisma, H., Machunik, M., Kakkonen, T., & Lebrun, J.-L. (2012). SWAN - Scientific Writing AssistaNt. A Tool for Helping Scholars to Write Reader-Friendly Manuscripts. In *Proceedings of the 13th Conference of the European Chapter of the Association for Computational Linguistics*, (pp. 20–24). Stroudsburg, PA: Association for Computational Linguistics.

Klein, D., Koltin, O., Peleg, M., Portnoy, I., & Greenbaum, D. (2017). Science and Law Separated by Impenetrable Language Barriers: Overcoming Impediments to Much Needed Interactions. *AJOB Neuroscience*, *8*(1), 37–39. doi:10.1080/21507740.2017.1285831

Laan, K. V. (2012). *The Insider's Guide to Technical Writing*. XML Press.

Lado, R. (1957). *Linguistics across Cultures*. University of Michigan.

Lamport, L. (1994). *LATEX: a document preparation system: user's guide and reference manual*. Addison-wesley.

Lave, J., & Wenger, E. (1991). *Situated learning: Legitimate peripheral participation*. Cambridge University Press. doi:10.1017/CBO9780511815355

MacDougall, M., Riley, S. C., Cameron, H. S., & McKinstry, B. (2008). Halos and Horns in the Assessment of Undergraduate Medical Students: A Consistency-Based Approach. *Journal of Applied Quantitative Methods*, *3*(2), 116–128.

Maiorana, F. A., & Mayer, H. F. (2018). How to avoid common errors in writing scientific manuscripts. *European Journal of Plastic Surgery*, *41*(5), 489–494. doi:10.100700238-018-1418-z

Markel, M. (2012). *Technical Communication* (10th ed.). Bedford/St. Martins.

Matthews, J. R., & Matthews, W. (2007). *Successful scientific writing: A step-by-step guide for the biological and medical sciences* (3rd ed.). Cambridge University Press. doi:10.1017/CBO9780511816185

McKercher, B., Law, R., Weber, K., Song, H., & Hsu, C. (2007). Why referees reject manuscripts. *Journal of Hospitality & Tourism Research (Washington, D.C.)*, *31*(4), 455–470. doi:10.1177/1096348007302355

Nesselhauf, N. (2003). The Use of Collocations by Advanced Learners of English and Some Implications for Teaching. *Applied Linguistics*, *24*(2), 223–242. doi:10.1093/applin/24.2.223

O'Connor, M. (2002). *Writing Successfully in Science*. E. & F.M. Spon. doi:10.4324/9780203478684

Ober, S. (2007). *Contemporary Business Communication* (7th ed.). Houghton Mifflin.

Orr, T., & Yamazaki, A. K. (2004). Twenty problems frequently found in English research papers authored by Japanese researchers. In *International Professional Communication Conference Proceedings* (pp. 23–35). 10.1109/IPCC.2004.1375270

Peat, J., Elliott, E., Baur, L., & Keena, V. (2002). *Scientific Writing: Easy When You Know How*. BMJ Books. doi:10.1002/9781118708019

Pierson, D. J. (2004). The top 10 reasons why manuscripts are not accepted for publication. *Respiratory Care*, *49*(10), 1246–1252. PMID:15447812

Richards, J. C., & Schmidt, R. (2002). *Longman's Language Teaching & Applied Linguistics* (3rd ed.). Pearson Education Limited.

Sageev, P., & Romanowski, C. J. (2001). A Message from Recent Engineering Graduates in the Workplace: Results of a Survey on Technical Communication Skills. *Journal of Engineering Education*, *90*(4), 685–693. doi:10.1002/j.2168-9830.2001.tb00660.x

Schein, M., Farndon, J. R., & Fingerhut, A. (2000). Why should a surgeon publish? *British Journal of Surgery*, *87*(1), 3–5. doi:10.1046/j.1365-2168.2000.01373.x PMID:10606903

Selinker, L. (1972). Interlanguage. *International Review of Applied Linguistics in Language Teaching*, *10*(3), 209–231.

Simionescu, M., & Simion, E. (2004). Scientific Lingua Franca and National Languages at the Crossroads. In P. Drenth & J. Schroots (Eds.), *ALLEA Biennial Yearbook. Critical Topics in Science and Scholarship* (pp. 129–133). ALLEA. Available at https://allea.org/wp-content/uploads/2016/02/Simionescu_-Lingua_Franca.pdf

Swales, J. M. (1997). English as *Tyrannosaurus rex*. *World Englishes*, *16*(3), 373–382. doi:10.1111/1467-971X.00071

Tang, R., & John, S. (1999). The 'I' in identity: Exploring writer identity in student academic writing through first person pronoun. *English for Specific Purposes*, *18*, 23–39. doi:10.1016/S0889-4906(99)00009-5

Thrower, P. (2012). *Eight reasons I rejected your article*. Elsevier online. Available at: http://www.elsevier.com/connect/8-reasons-i-rejected-your-article

Touchie, H. Y. (1986). Second language learning errors: Their types, causes, and treatment. *JALT Journal*, *8*(1), 75-80.

van Weijen, D. (2014). *How to overcome common obstacles to publishing in English*. Elsevier online. Available at: https://www.elsevier.com/authors-update/story/publishing-tips/how-to-overcome-common-obstacles-to-publishing-in-english

Ventola, E. (1992). Writing scientific English: Overcoming intercultural problems. *International Journal of Applied Linguistics*, *2*(2), 191–220. doi:10.1111/j.1473-4192.1992.tb00033.x

Ventola, E. (1994). Abstracts as an object of linguistic study. In S. Čmejrková, F. Daneš, & E. Havlová (Eds.), *Writing vs. Speaking: Language, Text, Discourse, Communication* (pp. 333–352). Gunter Narr.

Wee, R. (2009). Sources of errors: An interplay of interlingual influence and intralingual factors. *European Journal of Soil Science*, *11*(2), 349–359.

ADDITIONAL READING

Gennaro, S. (2016). Mistakes to avoid in scientific writing. *Journal of Nursing Scholarship*, *48*(5), 435.

Lebrun, J. L. (2011). *Scientific writing 2.0: A reader and writer's guide*. World Scientific.

Marina, V., & Snuviškiene, G. (2005). Error analysis of scientific papers written by non-native speakers of English. *Transport*, *20*(6), 274–279.

Provenzale, J. M. (2007). Ten principles to improve the likelihood of publication of a scientific manuscript. *AJR. American Journal of Roentgenology*, *188*(5), 1179–1182.

Swales, J. M., & Feak, C. B. (2004). *Academic writing for graduate students: Essential tasks and skills* (Vol. 1). University of Michigan Press.

KEY TERMS AND DEFINITIONS

Ambiguity: Ambiguity occurs when the meaning of an item can be interpreted in two or more ways.

Discourse Community: A discourse community consists of the people who read, write and/or use a particular set of discourses or text types.

Garden-Path Sentences: Sentences that lead most readers to parse the sentence and discover the incorrect interpretation first.

Generic Errors: Errors that occur when the language used would be considered inappropriate by members of the discourse community of the target genre.

Grammatical Errors: Errors that create expressions that would be considered incorrect according to the rules described in prescriptive grammar books.

Lexical Ambiguity: This type of ambiguity occurs when a term, usually a word, has two or more interpretations.

Lexical Errors: Errors that occur at the level of word choice and can be resolved by replacing the word with a more suitable choice.

Referential Ambiguity: This type of ambiguity occurs when a word or phrase can be interpreted to refer to more than one item.

Syntactic Ambiguity: Syntactic ambiguity or structural ambiguity occurs when a sentence can be interpreted in two or more ways due to the word order used in the sentence.

Vagueness: Vagueness is defined as the lack of precision or specific details.

Chapter 12
Reading the Narrative Structures in Educational Videos and Science Communication for the New Learner in the Age of Transhumanism

Basak Kalkan
Eskisehir Technical University, Turkey

ABSTRACT

When defining the new world, where knowledge is not independent of the one who possesses knowledge, it is considered imperative to develop new models for learning and science communication processes that are not simply confined to dichotomies such as tangible and intangible, or measurable and immeasurable. In the age of transhumanism, the concepts of production and consumption are freed from the shackles of absolute conceptualization and are defined instead with the concepts of producing consumer or consuming producer. In the science communication process, the new learner must be defined without being confined to absolute definitions on the new road map that is defined for the subject. The new learner is the actor, observer, adventurer, creative nomadic individual who cannot be trapped with absolutes in the post-modern era. This study seeks an answer to the question of whether or not a new content design is possible for the actor, observer, adventurer, creative nomadic individual in the face of the formlessness, ambiguity, and continuity of postmodernism.

DOI: 10.4018/978-1-7998-4534-8.ch012

INTRODUCTION

In the age of transhumanism, the first steps of a new learning design that produces problems and improves creativity will be taken by determining the definition and scope of design problems. Within the framework of theories of creativity, the three main components of creativity are fluidity, flexibility and originality. The digital natives of the age of transhumanism are fluid and flexible individuals with their multiple screen components. At this point, the issue at the forefront seems to be originality. Analyzing the design schools of the 21st century in terms of theory, methodology and ethics, Alain Findeli classifies the constraints of the designer as; the impact of the product, economic factors, a narrow philosophical anthropology, and a disregard for human and cognitive processes in an ergonomic environment. In the logical structure of the design process, the object of production is not the only way to transform the state of A into the state of B. Since the designer and the user are parts of the system, the design process is not static, but dynamic, and everchanging (Findeli, 2001). For this reason, it is seen that, in defining the new learner and developing the creative processes of this new learner, processes based on scenarios and narrative structures provide effective results for the student-designers to define previously undefined problems and transform abstract ideas into concrete codes. (Xu, Park, & Baek, 2011).

According to Kellogg (2008), the scenario and writing process utilizes three of our basic cognitive systems: memory, thinking and language. The individual who breaks free during the scenario process combines the problem with their own experiences and contextualizes the problem within their personal understanding. With writing, individuals discover what they really want to tell in their stories and what they really think (Miller, 2010). Writing is one of the most complex cognitive activities and involves numerous cognitive processes. The working memory, which supports cognitive processes may be reached when composing a text. The working memory is the cognitive field in which writing operations are performed. Working memory has attentional mechanisms for regulating when to activate and process knowledge necessary to compose a text. (Olive, 2012). For example, in the process of planning an article, ideas are taken from the long-term memory and reorganized if necessary. In addition, it is reported in studies that there is a relationship between memory and writing skills, that memory is an important determinant in the writing process, and that this is due to the memory capacities and individual differences in individuals (Bourke et al., 2014). Since creativity involves divergent thinking, the presentation of multi-faceted perspectives to learners will encourage learners to see relationships beyond their own ideas and discover new things. Thus, the learner will avoid being restricted to a certain thought, while at the same time being able to understand the relationships between different perspectives and ideas. Avoiding to structure the content in a strict way provides flexibility for tasks in the learning environment. Flexibility is an important factor in the development of creative thinking. Unstructured content will enable learners to discover new and different relationships. (as cited in Tezci & Gürol; 2003) In this context, providing answers to questions such as "What is a scenario?" "What is the role of the scenario in the science communication process?" is considered important in the transformation of practitioner-learners to designer-learners. As Wolfgang Jones puts it, the scenario itself is a design. It offers an advanced discipline for design, practice and conceptual framework. (Jones; 2001) Today, a scenario is used to describe processes that are structured in many different fields. In the context of this study, the use of the term scenario and the conceptualization of this term is a special form of writing that includes the events and dramatic moments of a story, waiting to be expressed in a visual form (Rea & Irving, 2004). Scenarios are tangible and flexible at the same time; they help developers manage the fluidity of their design states. Scenarios help developers manage many of the consequences of a particular design action, taking into account multiple views of the interaction, their variations, and the amount of details given. Finally, scenarios encourage communication about a problem among learners and include different socio-cultural and socio-economic indicators that will contribute to the design activity.

When we examine texts that are based on scenarios and narrative structures, another point to be considered would be the message format of the scenario. As Findeli points out, in the new logical structure of the design process, the object of production is not the only way to transform state A into state B. Since the designer and the user are parts of the system, the design process is not static, but dynamic, and undergoes infinite changes. Defining communication as the production and exchange of meanings, under the umbrella of the semiotic school of thought, Fiske says that the emphasis shifts to the text and how the text is being read. This means that, as Findeli states, a message is not something that goes from A to B. It is an element within a structured relationship, that includes other elements such as the producer/reader of external reality. Although producing and reading a text are not identical, they are parallel processes. It can be modeled as a triangle, where each arrow represents continuous interaction. The structure is not a static, but dynamic practice. (Fiske, 2003) In this context, it is necessary to create a new definition and a road map for the new learner in the age of transhumanism. According to Michel Foucault, a map is a tool of knowledge/power. This tool does not only define physical properties in the background. It is a tool that also reveals previously undefined relations of power.

SCIENCE COMMUNICATION IN THE AGE OF TRANSHUMANISM AND THE CHANGING LEARNER IDENTITY OF THE SUBJECT

In the new century, as we transition from the information age to the human age, the concepts of production and consumption are freed from absolute conceptualization, and are defined instead with the concepts of producing consumer or consuming producer. In this context, a new identity must be defined for the new learner and the message/messages must be structured according to this identity in the process of science communication. Answering the question "Who is the subject who defines and reconstructs the problem and has the self-confidence to change it", Calvino's definition of actor, observer, adventurer, creative nomadic individual finds its meaning with the new subject who is learning in the age of transhumanism. As stated by Calvino, who proposes a new city model for the 'utopian' travelers to embark on a playful and creative journey with their memories and desires as they please and to create spatial propositions for an architecture and urban environment that promotes a productive/creative/free urban life in the city of the past and present, that is governed by a rational, functional, guiding and controlling order, the new learner's identity must be allowed a productive/creative/free learning space. Another property that enables the realization of liberal and creative behaviors in the urban experience of the imagined places is the fact that these desired places are independent of location and the familiar 'safe' space that is rooted to the ground, which gives the feeling of a warm embrace. The dreams and the associations they evoke are, in a sense, homeless. They are far from the sense of belonging. (Bal, 2012) Calvino's actor, observer, adventurer, creative nomadic new learner is both the one who experiences, and the one who observes; and sometimes the one who only watches from the outside or the one who only experiences it by being in it.

In the era of transhumanism, the restructuring of the message is seen as an imperative for the new learner in the science communication process. In this framework, approaches that focus purely on communication theories are inadequate. Theories of learning and theories of communication working together in producing the message will ensure that the science communication process is carried out in a healthy and effective manner.

In redefining the learning processes for the new learner who experiences, observes and goes on adventures, multimedia tools and educational videos rise to prominence as new channels. These channels are also taken as models in different corporate structures, as well as in institutions that provide open and distance education in the process of knowledge transfer. Open learning processes come to the forefront as the prominent learning methodology for the age of transhumanism. The concepts of distance education, open education, web-based learning, computer-aided learning, e-learning, virtual education, which are the names of a new system for the instructors and learners of the new age, are used as the equivalent of the system. Although this conceptual differentiation is based on practical reasons, it opens up an important area of discussion in academic studies. Whereas the concept of openness defines flexibility and the right to choose in learning processes; the concept of distance defines an approach to education that is independent of time and space. (Aydın, 2011)

Since flexibility and the right to choose are key for the new learning subject in the age of transhumanism, open learning/education and open access are the prominent concepts of science communication in the new era.

Open education, which reverses the concepts of linear time and space, which are considered as barriers in education, sets the learning subject at its target as a system that is capable of reaching heterogeneous masses. Therefore, not only educational tools but also educational content become important. For any distance learning material, content takes up an important place in every aspect of the design. The content has the most important share in determining the success of any training program. As for the structuring of the content, the scenarioization process is the most important criterion for the success of the design. When we break down the process for the preparation of the training scenario within the framework of the Systems Approach, we can see that it involves 4 basic stages. They are as follows:

- Research
- Preparation
- Writing
- Assessment. (Yüksel, 2002)

In the open education system, the promise of multimedia learning is to increase the student's understanding. It uses words and visual materials to achieve this objective. Correct use of words and visual materials refers to the design process of messages. The cognitive theory of multimedia learning is based on three assumptions. These are: the binary code conjecture, the limited capacity conjecture, and the active learning conjecture. Binary code conjecture is based on visual and verbal presentation systems. According to this view, while animations and pictures are processed in the visual channel; while the auditory material is processed in the verbal channel. The limited capacity conjecture argues that the amount of processes that can take place for each piece of information is extremely limited. For instance, students can only remember 10 seconds of an animation at any given time. Therefore, according to the limited capacity conjecture, frequent repetitions are required in content design. The active learning conjecture argues that meaningful learning only occurs when the students engage in active cognitive processing. In the active learning conjecture, mental integration is ensured by organizing the verbal and visual materials. As a result of this active learning process, the learning outcomes are achieved as soon as problem solving occurs. (Mayer, 2003) In this framework, the schematic representation of the content to be designed for Mayer's meaningful learning is illustrated as follows. (Mayer, 2003)

Figure 1.

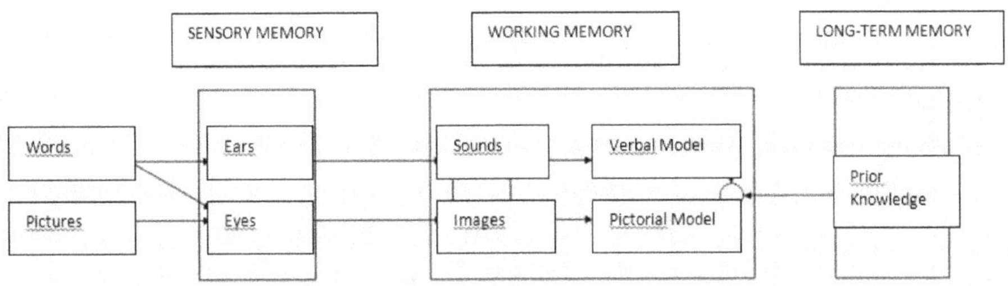

As Mayer stated, meaningful learning is learning that can solve problems. Meaningful learning, which is generally accepted as the aim of education, is learning by synthesizing, asking questions, questioning, and carrying that knowledge to social life. For this reason, the Elaboration Theory of Instruction, which is one of the new learner theories, is considered important in knowledge-centered message design. (Özerbaş, 2012) The objective of the Elaboration Theory of Instruction is to expand the Component Display Theory as previously developed by Merrill to the macro scale, i.e. to provide suggestions for the selection, organization, synthesizing, summarization of the instructional content. (Özonur& Tekdal, 2004)

In the design of detail sequencing, cognitive processing steps are used to analyze the structure of the information. In the Elaboration Theory, the details are sequenced according to the type of the content. The content types are categorized as conceptual, operational and theoretical. Accordingly, content design begins by asking questions, for conceptual content, what is the concept; for operational content, how to solve the operation and; for theoretical content, how to organize the hierarchical arrangement of information to go from simple to complex with wh- questions. To organize the hierarchical arrangement of information in a conceptual content, the mapping is performed from the general concepts to the special; in an operational content, from the simplest operation to the most complex. If there is prerequisite learning with regards to the content, prerequisite information must be placed ahead of the content. Then we can proceed with the summarization stage.

Synthesizers and analogies are the important steps of the Detailing Theory. Synthesizers are used to absorb the learner. It is a macro strategy component that enables the learner to draw associations between the information that the learner knows beforehand and new information. Analogies can be categorized into four types. They are as follows:

- Simple analogy
- Analogies in story form
- Gamified analogies
- Analogies created by pictures.

Analogies are also considered important in content design as memory supplements. (Reigeluth, 1998)

At this point, the important thing to achieve is allowing the audience to choose the microstructure they want within the learner, by creating the message - which is a scenario, for multimedia tools - with a modular design. In the production process of instructional scenarios designed according to the Elaboration Theory of Instruction, it is important to work in conjunction with the Detailing Theory and Mayer's cognitive learning theory for multimedia environments. In this context, production processes consist of the following steps:

Figure 2.

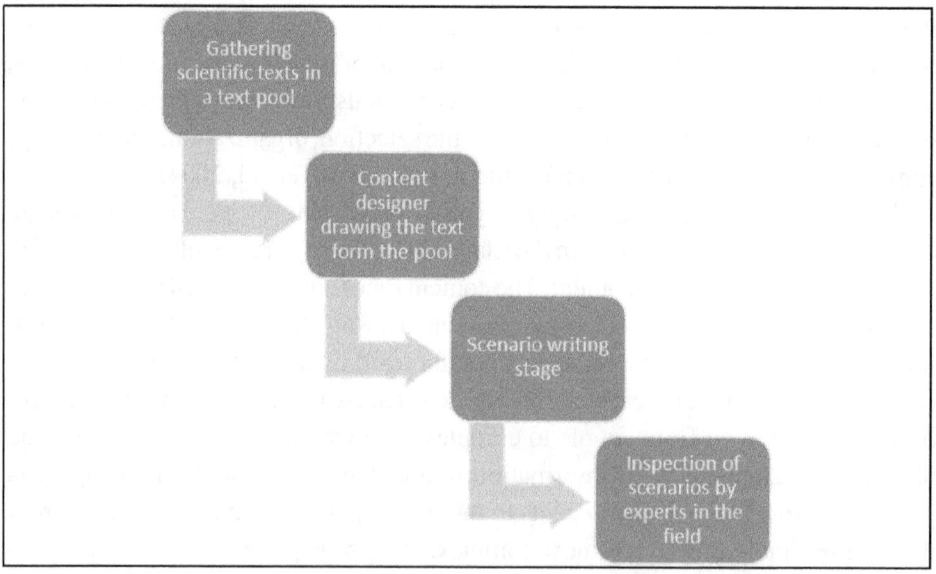

According to the Education Detailing Theory, the scenario writing stage goes through the following steps:

Figure 3.

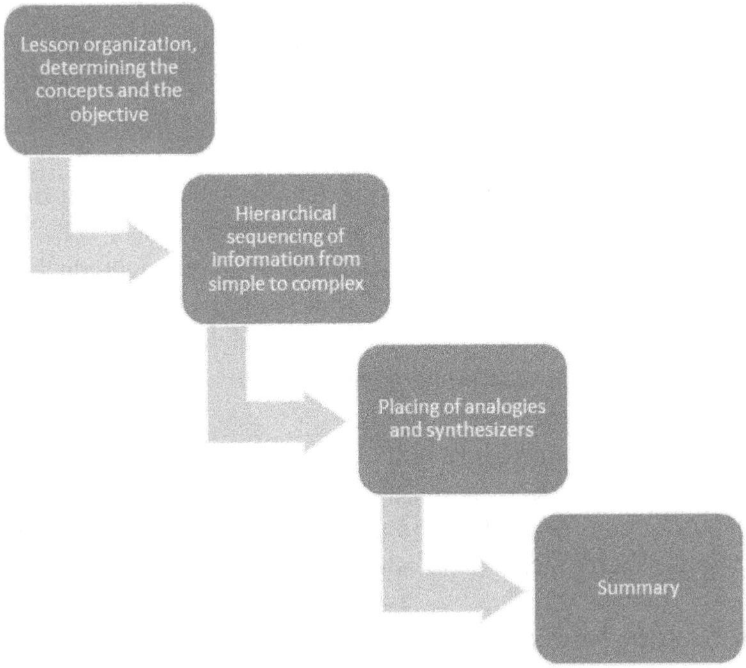

In the era of transhumanism, the restructuring of the message is seen as an imperative for the new learner in the science communication process. In this context, while encoding the message the new learner, who is defined as the designing/producing learner, must be included in the process. Calvino's actor, observer, adventurer, creative nomadic new learner is both the one who experiences, and the one who observes; and sometimes the one who only watches from the outside or the one who only experiences it by being in it. Within this framework, we need to identify the new learner, who is characterized by dominant creative attributes, as a designer.

Various studies have been carried out since the 1960s in order to understand how the designer performs the design action and their frame of mind during this process. An important part of these studies is related to the definition and scope of design problems. As a result of his research, Rittel has developed the definition of "wicked problem" in the 1960s. (Ekmekçioğlu & Özkaraman; 2017) Buchanan asserts, as the basis of these creative accidents, the fact that designers work with poorly formulated social system problems surrounded by complicated data. (Buchanan, 1992: 15). Cross (2006: 7) states that, based on the fact that there is no absolute truth, design problems do not inherently contain all the information that is necessary for a solution. Stating that designers should have the confidence to define, reconstruct and change the given problem in order to handle problems that are not fully defined, Cross describes the designer's perspective under five items: 1. Designers deal with problems that are not fully defined 2. Their way of problem solving is solution-oriented 3. They have a constructive way of thinking 4. They use codes to transform abstract needs into physical objects 5. They use these codes to "read" and "write" in the "language of the object" (as cited in Ekmekçioğlu& Özkaraman; 2017).

In the age of transhumanism, the designer's perspective given by Cross is considered important in terms of understanding the explanatory subject of the message in the process of science communication. The new designing/producing learner; is a subject that defines its abstract needs, with physical codes, who is concerned with problems that are not fully defined, has a solution-oriented approach and a structurally constructive way of thinking, and because of this attribute, a modular message structure is essential.

In this context, while configuring the new message in the process of science communication, a new design needs to be developed without getting stuck in the absolute definitions of the learner/instructor, practitioner/designer within the framework of the source and receiver relationship.

Content Design for New Learner in the Age of Transhumanism

Science communication is the practice of informing, educating, sharing curiosity and raising awareness in science related issues. It can be seen that new technological developments have changed the lives of the societies in the current era. Every newly discovered technological development and economic events deeply affect societies, and therefore individuals. For instance, the discovery of the steam engine led to the Industrial Revolution, and in turn the Industrial Revolution brought about the working class as a new class, while the socialist doctrines of the working class introduced Marx's ideas to the world of thought. In the twentieth century, Marx's theses and anti-theses formed against them introduced new forms of government to this century's communities. Economic and political events as well as technological developments have a deep effect on the lives of communities. The economic and political turmoil, which started especially in the first quarter of the twentieth century, has caused significant changes in the world's political arena, while deeply affecting the lives of communities, and therefore individuals.

Another dynamic that changes and improves the lives of societies is, undoubtedly, communication. While technology helped develop the communication tools, it furthered its own scientific advancement with the help of communication. The invention of the printing press marked the transition to printed communication and helped the advancement of science and technology thanks to the books that are printed and distributed all over the world. This interaction between technology and communication has now completely turned the 21st century into an age of communication.

As for communication as a field of science; following the 1929 Global Economic Depression, the necessity to explain what was going on to the society and the importance of communication was recognized and communication took its place in the world of science as an interdisciplinary field in the background. Of course, communication did not enter the social and cultural life after twenty centuries. It is a phenomenon that has existed since the existence of the communicating person. However, in the recent years, more importance has been attached to the effects of communication on the society and the cultural life.

In order to evaluate the concepts of communication, society and culture, firstly, the concepts of society and culture must be taken into consideration in terms of terminology. While it is possible to explain the relationship between the concepts of communication and culture based on the approach: "communication involves both sharing things that are new and sharing things that are common", the interaction between communication, society and culture has been the most discussed topic in recent years (Lundby & Ronning, 2002). It can be seen that, communication has been influenced by and has benefited from traditional culture, developing and changing the culture as it developed, and that societies have become engaged in the construction of a single culture in the maze of multiculturalism during the new age.

In order to explain the concept of society, it is necessary to answer the question "what is coexistence?" According to Aristotle, a human is a "zoon politicon (social animal)". Coexistence is not unique to humans. Many animals also live in flocks. However, the sociability that exists in animals is instinctive. On the other hand, it seems that sociality is the product of the mind, since people are aware that they cannot live on their own. According to Aristotle's definition, "having a mind and the power of language, which is the expression of the mind" is the most important attribute that distinguishes man from animal (Öktem, 1995). The most important factor that urges people to live together is psychology. First of all, the individuals who come together as part of a whole all have a certain psychological structure. When the events that occur at the level of consciousness and at lower levels, in the internal communication of the person with themselves, are taken into consideration by isolating the person from the environment and the society, the idealized loneliness of the individual and their position with regards to other people are subject to philosophies that emphasize psychology. In Sigmund Freud's psychoanalysis, the communication of the person with themselves is considered as part of a spiritual mechanism. Communication between the "id", which is the repository of motivational information, and the "super ego" which is completely formed by the information from the external world, plays the most important roles as the basis of human behavior. Cari Gustav Jung, the founder of analytical psychology, analyzed the characters of introverted and extroverted people as the products of their internal communication and pointed out that there is an accumulation of collective information, inherited by individuals from their predecessors, at a subconscious event level. Psychology plays a critical role in the communication of individuals with one another within the society. However, it is seen that neither sociology, nor psychology alone is effective in explaining this communication with one another. Psychology and sociology act in parallel. As stated by Karl Jasper, the internal communication of the individual develops with their own history. Jasper says, "There is no abstract individual. They exist only through others …" (Evliyaoğlu, 1987).

The most important common feature of society and culture is that they are integrative and subject to change. Both society and culture develop and change over time due to both political and economic factors. Culture is particularly affected, changed and developed by the culture around it. Moreover, economic factors affect social structures and cultural identities. Communication is a very important actor both in the internal structure of societies - in the transfer of cultural identity to the individual, in the imposition of culture on the individual - and in the external world.

Communication, on the other hand, describes human activity. Where there is human activity, there is communication. Communication is the basic condition of human life and relationships. According to the definition of Schiller, communication involves much more than messages and circuits through which messages flow. Communication defines social reality, in this way affects the organization of work and affects the character of technology, the curriculum of the educational system, the use of spare time. Communication is a necessity. Necessity brings about purposeful human activity. Therefore, the purpose of human activity becomes communication. Even when it seems pointless, it definitely has a purpose.(Erdoğan, 1997)

Communication has always been affected by the socio-economic and political developments of societies, and technology has brought along new communication channels in cases where communication channels prove insufficient. Not only communication channels have changed in the process of social change, but also the ideological and cultural changes of societies have changed and developed the definitions of communication. As of today, communication, as a field of science, is divided into different interdisciplinary fields called organizational communication, political communication and science communication.

Three important elements come to the forefront in the message receiving and sending system, which is one of the predominant approaches in communication. Sending and receiving messages is the system, and role of communication in the system. According to the component "sending and receiving messages", the message, in its broadest sense, is the transmitted communication. In this sense, communication occurs between the sender and the recipient and is a process that connects them. These processes include creating, sending and receiving messages. The system generally consists of four elements: objects (that make up the system, can be concrete or abstract according to the nature of the system), features (the properties of the system and the objects), internal relationships (refers to the relationships between objects, common impacts and boundaries), environment (the system cannot exist on its own, it is surrounded by an environment. The environment is made up of other systems in which the system exists and maintains its existence). Finally, the role of communication in the system is expressed as the sum of the activities of the renewable parts that work together to create forms of relationship between objects in space and time. According to the prevalent understanding, communication serves four main purposes: entertaining, giving information, persuading, alerting. On the other hand, approaches that consider communication as a social structure state the role of communication as keeping the society in a functional balance. (Erdoğan, 2005).

In this context, the message production for the new recipient in science communication must be designed in consideration of the characteristics of the system, in addition to internal relations and the environment.

Science communication is seen as an important area in the formation of a collective consciousness and new life practices of a modern society, as well as in organizing the society. While the liberation of the urban subject is built with social involvement, the language of scientific communication sheds light on the future of the society. The evaluation of science communication in a broad framework beyond inter-object relations, from subject and inter-subject relations to social facts, will help provide concrete suggestions in structuring science communication with a view to achieve universal benefit.

As we transition from modern society to the post-modern society, the knowledge of the current era plays an organizational role. As Foucault puts it, knowledge is not disciplinary. For this reason, the target is not a single piece of information, but consists of the combination of many pieces of information (Urhan, 2011). In the construction of this new process, communication is the most important network of the change process. Habermas' communicative rationality, which points to intersubjectivity, also points to communities as a very important mass media tool in the construction of the public sphere. This new communication model is a new construction process based on the relationship between the subject and the subject; the subject and the object; the object and the object.

This social theory progresses towards the goal of human liberation while establishing a comprehensive universal moral framework. This framework is based on universal benefit, the goal of mutual understanding and the argument that human beings have communicative competence. In social life, when the readability level of knowledge is evaluated within the framework of universal benefit, it is considered important in terms of human liberation, and defines a field of communicative rationality and communicative action in which people do not consider themselves to be the only subject.

Communicative rationality, in which people do not see themselves as the only subject, defines a collective language of science. This collective language of science is in turn defined within a system of signs and indicators. The observer, actor, creative nomadic subject of the modern world is exposed to different signs at every moment of their daily lives. These signs are common identity indicators that describe scientific knowledge, transfer scientific knowledge, and establish communication between knowledge and the people with the meanings attributed to them, forming the language of scientific communication for the purpose of facilitating their daily lives. In the process of science communication, it is necessary to structure the subject's relationship with the other subjects and objects and to prepare the indicators in this framework with a subject-focused approach. In terms of educational videos that science communication uses to reach large audiences, it is considered important to evaluate the message structure within the framework of the theories of learning and the theories of communication. For the new subject in the age of transhumanism, message design in science communication has to go beyond knowledge transfer. Above and beyond the accessibility of information, the availability of information is what's important for the observer, nomadic new subject. In this context, storification is used as an important tool in message design.

STORYTELLING IN THE SCIENTIFIC COMMUNICATION PROCESS FOR THE NEW LEARNER IN THE AGE OF TRANSHUMANISM: MODEL MESSAGE DESIGN FOR EDUCATIONAL VIDEOS

The culturally dominant view of the popularization of science is based on a two-stage model: first, scientists develop actual knowledge; second, popularizers distribute the streamlined versions to the public. Popularization serves as a resource in public discourse, offering a repertoire of rhetorical devices to interpret science and a powerful tool to maintain social hierarchy. (Wynne, 2008)

While storytelling itself generally has negative connotations in science, theories of storytelling and narrative should not be ignored when transmitting scientific knowledge to non-expert audiences. Narratives offer more understanding, interest and engagement. Non-experts receive most of their scientific knowledge from mass media tools, which already tend to utilize narrative forms. The narratives are also intrinsically persuasive, which provides scientists with tactics to persuade masses that are resistant against science communicators. Scientific communication aims to provide abstract truths that apply across a specified range of conditions. One can then use these abstract truths to generalize a particular case and ideally provide a certain degree of predictive power related to a particular situation. Narrative communication provides scientific communication a specific case in order to reveal what the general truths must be. (Dahlstrom, 2014).

The differences between scientific communication and narrative communication have led to an understanding that claims, that not only these forms of communication are opposite, but they also represent two different ways of cognitive understanding. The paradigmatic path controls the encoding of science-based evidence, while the narrative path controls the encoding of case-based samples, leading to significant differences in understanding content, and also understanding the content based on the path used to process such content. (Dahlstrom, 2014).

It is important to incorporate theories of learning and theories of communication together while encoding message structures in the scientific communication process. While designing a scenario for a multimedia tool, it can be seen that a template of the learning process is also created. When a learning process is examined; the progress is revealed to take place in five steps: attracting attention, discovery, explanation, elaboration (transfer) and evaluation. The scenario writing process also consists of the conflict, development, climax and conclusion parts. When two different processes are evaluated, attracting attention, which is accepted as the first step of the learning process, corresponds to the conflict part of the scenario as a narrative structure. Conflict draws attention to the text and describes a striking and difficult situation. Discovery and explanation steps correspond to the development part in the scenario process; while elaboration (transfer) corresponds to the climax part and evaluation corresponds to the conclusion part. While encoding the message in the scientific communication process, it is important to encode the information in alignment with folklore without ignoring the relevant ethical principles, in order to establish an interaction between scientists and people. In this context, let's try to understand the differences of narrative communication in terms of the transfer of knowledge with a model.

Example Scenario: (TRT OKUL - Prolog - Episode 1 (Postmodernism/Time)

Table 1.

FADE-IN **STUDIO** STANDING NEAR THE CHAIR. EXAMINES AND TOUCHES THE TEXTURES OF THE CHAIR. TURNS AROUND AND SITS DOWN ON THE CHAIR (sits down quickly.) THE ACTOR IS HOLDING A TUMBLER. HE LEAVES IT ON THE FLOOR BEFORE SITTING ON THE CHAIR. THE TUMBLER IS LEFT HANGING IN MIDAIR. A TUMBLER FULL OF WATER ON A COFFEE TABLE. IN THIS SECTION WE ONLY SEE THE TUMBLER. AS THE CAMERA ZOOMS IN, THE TUMBLER STARTS TO SHAKE. (sound off) CHAIR PLAN TALKS WHILE HE'S SITTING ON A CHAIR, LOOKING AT THE CAMERA. THE DOUBLE CAMERA FLASH CUTS... SHIFTS TO A LOUNGING POSITION ON THE CHAIR. **THE NAME OF THE ROSE-UMBERTO ECO PG. 517......** **THE MAZE WILL BE A GRAPHIC OR MADE OF WOODEN MATERIAL!!!** **THE ACTORS TALKS ON HIS CHAIR..** (WATER FOAM WILL FALL DOWN ON THE BONSAI TREE) **TRANSITION......STUDIO** **OUR CHARACTER TALKS AFTER TURNING HIS HEAD FROM A LOW ANGLE.** Only sits and talks, looking at the camera. While he's talking, the follow spot slides off of him. STARES WITHOUT BLINKING AND MOVING. (SOUND-OFF) LIGHT EFFECTS CONTINUE. THE TEXT IS DISPLAYED ON THE SCREEN IN A GRAPHICAL FORM. EXTREME CLOSE-UP.(WIDE LENS) **TRANSITION** **LOMOKINO IMAGES. A TREE THAT IS SPOTTED FROM THE WINDOW.** **ANOTHER CHARACTER WHO IS WALKING THROUGH THE WOODS, HOLDING A SIGNAL FLARE..** SITS ON THE BACK OF THE CHAIR, TALKING. TYPOGRAPHIC LETTERS WILL APPEAR. THE SOUND WILL BE OFF FOR THIS SECTION. MOTION TYPOGRAPHY LETTERS WILL FORM THE MAZE. TRACK-OUT WE SEE THE TUMBLER FALLING DOWN FROM THE COFFEE TABLE TO THE FLOOR IN HIGH FRAME RATE. **FADE-OUT**	**ACTOR:** Now, if I start talking about texts, I will be faced with the question "What kind of texts?". If I start talking about literature, you will ask me"Doesn't that include texts?". Or maybe you won't ask me any questions but simply wonder "What the hell is this guy going on about?" It's possible. The only thing I care about is searching for time, in a world of literature with capital letters, and a world of text with tiny letters. That's one too many words for a man who is looking for time in a madman's diary. All I'm trying to do is talk a little about time and a little about texts, through the diaries of a lunatic who has lost time or trying to rediscover what wasn't there in the first place. My confusion between the concepts of literature and text is the reason for everything, even this program. If I can get it done today, and unfortunately there is no precedent, I want to pull the texts out of their endless maze. Our concern is to analyze a discussion, that is concerned with not the writer, but the reader, at an indefinite time. For this, we can start by ignoring the time, the place, the subject, the heroes, and even the narrator of the text. Speaking of today; today, tomorrow or yesterday, no matter where you look at it; we have to put aside time as well. My concern is to take you to a moment, right here, right now, in this chair, in the folds and buttons of this chair. I'm getting ready to pull you out now, whenever you're ready. **ACTOR:** Now let's go to a moment that is independent of time. Post-modern texts and time. Where to start? Both are quite challenging. Superstructure, parody and irony are the first things that come to mind. The disorder formed by the superstructure, parody and irony taken from the supertext. As for Time, it is even more difficult to define that. According to some, it is the infinite period during which all living things are connected to each other; and according to others it doesn't even exist. And I'm here trying to establish some kind of order within disorder. **ACTOR:** DAY SIX.... MIDNIGHT. We came down at midnight for prayer. In this last part of the night, almost in the first part of the new day, the sky was still cloudy. As I passed through the courtyard, moisture was penetrating into my aching bones and down to my marrow after a restless slumber...... ...... They had just begun chanting the hymns **ACTOR:** ITALO CALVINO - INVISIBLE CITIES PG. 117 Now many seasons of abundance have filled the granaries. The rivers in flood have borne forests of beams to support the bronze roofs of temples and palaces .... The Great Khan contemplates an empire covered with cities that weigh upon the earth and upon mankind, crammed with wealth and traffic, overladen with ornaments and offices, complicated with mechanisms and hierarchies, swollen, tense, ponderous. **ACTOR:** I'm reading a medieval novel without rejecting canonical hours, and I'm trying to understand today's urban life, on an intellectual journey from Marco Polo to Kublai Khan. Moreover, I try to do this without getting stuck in absolute time. The break of dawn Eco determined and even presented to the reader as a guide in the Name of the Rose, is it midnight or morning; is time Calvino's Great Khan's contemplation of his empire from past to the present? Time that creates or is created without getting stuck in time. To put it much more simply, the time of the author, the time of the work and of course the time of you. **ACTOR:** **OĞUZ ATAY - THE DISCONNECTED** "Which fault of yours did they give you the opportunity to correct? Did they let you know that your journey was coming to an end before you arrived at the last stop? They just suddenly went: "This is it!" If you only knew, you would look at the stations with full attention; you wouldn't miss a single tree, a single patch of sky you can see out the window. You would watch your shadow in all the puddles. And to think they would say "we had informed you before." "We had told you in general, that everything has to end. They get old, as you live them, and make you old." **ACTOR:** The author's time, the work's time or your time. As I go with Oğuz Atay, who carried you to a finite time but lost the moment, I again question the postmodern novel and the order within disorder, this time more vigorously. How much does Eco's time in the reader's guide, that rejects canonical hours, coincide with real time? Or as for Calvino's work, are there any other signs of time other than the traces it leaves on places. Or is the lost moment the only thing that is objectified in Oğuz Atay's The Disconnected? The answer to these questions points to the exit of the maze. They left the last sentence of the novel open, for you. For me? I'm already stuck desperately in time.

The example narrative structure is the first episode script of the production called Prolog, which is a literature program made by the Anadolu University Television Production Center for the TRT School channel that broadcasts under the Turkish Radio Television Corporation. The aim of the program is to introduce literary movements to the audience in alignment with literary sciences. In this program, which is envisioned as 13 episodes, a concept corresponding to each movement was determined and the program was designed and enriched with examples from the world and Turkish literature within the framework of this concept. The purpose of the design process is to establish a narrative structure that is not detached from the context in the reality of science communication. In the program where the concept of time is explained within the framework of postmodernism, the basic characteristics of the post-modern literature are explained through the concept of time. The point to be considered here is that, while structuring the process of science-human interaction and establishing the narrative structure with the individual, it must be ensured that the visual and verbal codes used are included in the individual's knowledge map, and the story of the individual and the story of science must be combined in the context of postmodernism. A second point to consider is that the text must be encoded within the framework of the theories of learning. After determining the content, discovering the interest, opening the program and identifying the difficult problem;

"Now let's go to a moment that is independent of time. Post-modern texts and time. Where to start? Both are quite challenging. Superstructure, parody and irony are the first things that come to mind."

The program was designed by determining the concepts and the purpose. In the progressive structure of the program, the messages were coded by organizing the examples from the world and Turkish literature from the easiest to the most complex structure and these were supplemented with graphical elements. In graphical elements, typographic visual elements were preferred, avoiding visual noise, considering the weight of the scientific structure. Analogies and synthesizers are the basic elements of the process from the first moment up to the last moment of design. The tumbler hanging in the air, the tumbler falling off of the coffee table, the selection of the bonsai plant that symbolizes the concepts of growth and birth, were all used as analogies between time and postmodernism. Here, the concept of time was also used as a synthesizer and the link between postmodernism and artistic works was established through the concept of time.

CONCLUSION

Calvino offers new propositions with regard to the new learner's identity in order for the 'utopian' travelers to embark on a playful and creative journey with their memories and desires as they please and to create spatial propositions for an urban environment that promotes a productive/creative/free urban life in the city of the past and present, that is governed by a rational, functional, guiding and controlling order. In the science communication process, the new learner must be allowed a productive/creative/free learning space. Calvino's actor, observer, adventurer, creative nomadic new learner is both the one who experiences, and the one who observes; and sometimes the one who only watches from the outside or the one who only experiences it by being in it. The new designing/producing learner is a subject that defines its abstract needs with physical codes, who is concerned with problems that are not fully defined, has a solution-oriented approach and a structurally constructive way of thinking.

In this framework, the message in the science communication process should be redesigned for this new learner, and encoded by taking into account the narrative structures for the actor, observer, adventurer, creative nomad, which are in a close relationship with the context, together with the learning processes. The richness of the content in these channels and the channels available to you is important for the encoding of the message for the nomadic subject who is independent of space in the age of transhumanism. The main points of consideration while encoding the message in the process of science communication for the new learner within the framework of narrative structures, can be listed as follows:

- Determine the constraints of the learner well, do not leave the information alone without context
- Structure the assumption that will catch the attention of the viewer/ reader well. Difficult gathers attention, establish the difficult and striking structure during the introduction.
- Decide on your format. This model can be theatrical storytelling, a documentary film, or a format supplemented by interviews, as given in the example.
- Don't forget that you can use auditory and visual elements. Feel free to use ambient sounds, graphical elements and images to support your message.
- One of the most effective elements of a video is the transitions. The power of fluidity, flexibility and originality, which are the basic elements of a design process, echo in the rhythm, tempo and flow of a video. The emphasis felt while watching, the magnitude of this emphasis, and the changes between the magnitudes of this emphasis will reveal the strength of the message.

Access to knowledge is a basic human right. In the age of transhumanism, a new content encoded with narrative structures for the new learner will carry science to point where it is more freely accessible.

REFERENCES

Aydın, C. H. (2011). *Open and Distance Learning Candidates' Perspective*. Pegem Academy.

Bal, B. (2012). Imaginary Atlases: "Timeless Journeys" with Constant's "New Babylon" and Calvino's "Invisible Cities." *Journal of SKOP, 2*. Retrieved from https://www.e-skop.com/skopdergi/dussel-atlaslar-constantin-new-babylonu-ve-calvinonun-gorunmez-kentleri-ile-zaman-otesi-yolculuklar/581

Buchanan, R. (1992). *Wicked Problems in Design Thinking*. doi:10.2307/1511637

Dahlstrom, M. F. (2014). Using Narratives And Storytelling To Communicate Science With Nonexpert Audiences. *Proceedings of the National Academy of Sciences of the United States of America, 111*(Supplement_4), 13614–13620. Advance online publication. doi:10.1073/pnas.1320645111 PMID:25225368

Ekmekçioğlu, D., & Özkaraman, M. (2017). Defining Self-Organizing Systems as a Design Problem. *Online Journal of Art and Design, 5*(4), 89-101. Retrieved from http://www.adjournal.net/articles/54/546.pdf

Erdoğan, İ. (1997). *Communication, Domination, Introduction to Struggle*. Imge Publishing.

Erdoğan, İ. (2005). *Understanding Communication*. Erk Publications.

Evliyaoğlu, G. (1987). *Communication Psychology, Psychological Communication*. Journalists Association Publications.

Findeli, A. (2001). *Rethinking Design Education for the 21st Century: Theoretical*. Methodological, and Ethical Discussion. Retrieved from https://www.mitpressjournals.org/doi/pdf/10.1162/07479360152103796

Fiske, J. (2003). İletişim Çalışmalarına Giriş. Ankara: Science and Art.

Hilgartner, S. (1990). The Dominant View of Popularization: Conceptual Problems, Political Uses. *Social Studies of Science, 20*(3), 519–539. doi:10.1177/030631290020003006

Illingworth, S., & Allen, G. (2016). *Effective Science Communication*. Bristol, UK: IOP Publishing. Retrieved from http://webcache.googleusercontent.com/search?q=cache:reKcQk72mzkJ:library.iyte.edu.tr/dosya/kitap/efective.pdf+&cd=1&hl=en&ct=clnk&gl=tr

Isman, A. (2011). *Distance Education*. Pegem Akademi.

Jonas, W. (2001). A Scenario for Design. *Design Issues, 17*(2), 64-84. Retrieved from https://www.mitpressjournals.org/doi/pdf/10.1162/07479360152383796

Kellogg, R. T. (2008). Training Writing Skills: A Cognitive Developmental Perspective. *Journal of Writing Research, 1*(1), 1-26. Retrieved from http://citeseerx.ist.psu.edu/viewdoc/download?doi=10.1.1.569.8025&rep=rep1&type=pdf

Lundby, K., & Ronning, H. (2002). Media, Culture, Communication: Interpretation of Modernity Through Media Culture. In Media, Culture, Politics. Ankara: Alp Press.

Mayer, R. E. (2003). *The Promise of Multimedia Learning: Using the Same Instructional Design Methods Across Different Media, Learning and Instruction*. Retrieved from https://www.journals.elsevier.com/learning-and-instruction

Miller, L. C. (2010). *Make me A STORY*. Stenhouse Publishers. Retrieved from https://books.google.com.tr/books?hl=tr&lr=&id=dIINvn2RGUAC&oi=fnd&pg=PT8&dq=miller+2010+writing&ots=LIIQ6IGhLu&sig=bfQEpsNyIZ59AQLsFr_6SbEX1yE&redir_esc=y#v=onepage&q=miller%202010%20writing&f=false

Oktem, N. (1995). *State and Legal Philosophical Movements with the Data of Sociology and Philosophy*. Der Published.

Olive, T. (2012). Working Memory in Writing. Ed. Virginia Wise Berninger. In *Past, Present and Future Contributions of Cognitive Writing Research to Cognitive Psychology*. New York: Psychology Press Taylor & Francis Group. Retrieved from https://books.google.com.tr/books?id=MoaTPgbuAccC&pg=PA485&lpg=PA485&dq=olive+2004+writing&source=bl&ots=uYZgIiAI60&sig=okR1hmIcFcTh3pKIknS6KgcCdG8&hl=tr&sa=X&ved=0ahUKEwjZ1pPJaTYAhWRIOwKHZELBvIQ6AEITjAF#v=onepage&q=olive%202004%20writing&f=false

Ozerbaş, M.A. (2012). Öğretimi Ayrıntılama Teorisi, Elaboration Theory. *GEFAD/GUJGEF, 32*(2).

Ozonur, M., & Tekdal, M. (2004). *The Effect of Web Based Asynchronous Distance Education Application Designed Based on the Detailed Theory of Teaching on Academic Achievement of University Students*. XIII. National Educational Sciences Congress, Inonu Universty, Malatya, Turkey.

Rea, P. & Irving, D. (2004). *Cinema and Short Films on Video- 1, Pre- Production* (S. Taylaner, Trans.). İstanbul: Es.

Reigeluth, C. M. (1998). *Elaboration Theory: Scope and Sequence Decisions for Quality Instruction.* Retrieved from https://www.reigeluth.net/pubsinsttheor

Tezci, E., & Gürol, A. (2003). Constructivist Instructional Design and Creativity. *TOJET, 2*(1), 50-55. Retrieved from http://www.tojet.net/articles/v2i1/218.pdf

Urhan, V. (2011). *Pedigree of M. Foucault and Knowledge / Power Relationship.* Retrieved from http://www.uludag.edu.tr/~kaygi/dergi/articles/9_99-118.pdf

Wynne, B. (2008). Public participation in science and technology: Performing and Obscuring a Political-Conceptual Category Mistake. *East Asian Science, Technology and Society, 1*(1), 99–110. doi:10.100712280-007-9004-7

Xu, Y., Park, H., & Baek, Y. (2011). A new approach toward digital storytelling: An activity focused on writing self-efficacy in a virtual learning environment. *Educational Technology & Society, 14*(4), 181-191. Retrieved from. http://www.ifets.info/journals/14_4/ets_14_4.pdf

Yüksel, C. (2002). Some Thoughts on the Education Scenario - A Systems Approach. *Journal of Selcuk Communcation, 2*, 3.

ADDITIONAL READING

Benedict, R. (2000). *Culture Patterns* (M. Topal, Trans.). Öteki Publishing.

Guvenç, B. (2005). *Human and Culture*. Remzi Publishing.

Janh, M. (2015). *Narratology: Narrative Theory Handbook* (B. Derviscemaloğlu, Trans.). Dergah Publishing.

Mayer, R. E. (2014). Cognitive Theory of Multimedia Learning. In The Cambridge Handbook of Multimedia Learning. Retrieved from DOI: doi:10.1017/CBO9781139547369.005

Ozgen, M. (2006). *Journalist's Ethical Identity*. SET-Systems Tercuman Publishing.

Yüksel, A. H. (2003). Communication Concept and Definition. In Demiray (ed.), General Communication for Vocational Schools. Ankara: PegemA Publishing.

Chapter 13
The Use of Animation and Storytelling in Science Communication

Cigdem Tas Alicenap
Anadolu University, Turkey

ABSTRACT

Science communication has brought new teaching/learning environments forward. Various digital media have provided appropriate platforms. Learning processes for everyone have been differentiated and thrived on newly developed technologies and digital media. Animation is among these to provide an alternative to science communication methods. Especially animations developed with storytelling can become very effective in learning processes. In cartoon movies, certain knowledge on different fields has been explained by turning them into scenarios. It's possible to see examples of cartoons in this concept in Turkey and all around the world. Storytelling (scenario) and animation dynamics make knowledge more impressive and long-lasting. Educational cartoon movies produced for children, which aren't in formal and organized education but younger than the school age, have been produced in a widespread manner. This study is focused on transferring scientific knowledge by using animation products. In this context, educational cartoon movies will be examined through storytelling and animation.

DOI: 10.4018/978-1-7998-4534-8.ch013

INTRODUCTION

With the changing world order and developing technologies, all disciplines has begun to use digital platforms excessively in order to transfer knowledge to the individuals. Especially in 2020, due to the Coronavirus disease (COVID-19) pandemic which has affected the whole world, numerous scientific fields such as education have designed materials for digital platforms and have begun to benefit the effectiveness of visual communication design more than ever. It is not surprising to see an increase in the need for visual communication channels in the field of education. "*The researches that were performed on students' learning processes have shown that visual perception is the most developed sense of human beings and it is an important way for learning*" (Sekular and Blake, 1985, as cited in McClean, et.al., 2015, p. 170). Visualization helps in understanding complex processes.

The Use of Animation in Knowledge Transfer

Being on the rise in the recent years, animation with its technical features is not only an important tool for entertainment industry (cartoon movies, movies, series, video games, etc.) but also for the field of education. Animation can especially be used to teach the processes which are difficult and complex to comprehend. The best examples of this fact can be found in the field of medicine. Medical movies prepared by using animation techniques make understanding the complex and unseeable medical procedures easier for both physicians and for medical students. Animation is important since it enables bringing animation, voice and text together to transfer a complicated process to the target audience apprehensibly. In this process, a well-designed educational multimedia materials play a crucial role in fast and long lasting learning experience of individuals by seeing and hearing. Different literature researches have shown that moving images and audio materials have positive effects on teaching and learning behaviors.

Two theories on learning have a specific importance on understanding the importance of animation design in terms of teaching and learning. These are Cognitive Load Theory and The Cognitive Theory of Multimedia Learning (Ruiz, et. al., 2009, p.839). "*Cognitive Load Theory puts forward that working memory can hold a limited amount of information in a limited period of time. When this limit is exceeded, a cognitive load emerges and learning process gets impeded... In terms of cognitive load, one of the goals of teaching is to decrease extraneous cognitive load while increasing germane cognitive load. Therefore, the working memory will be freed to perform learning and understanding. The Cognitive Theory of Multimedia Learning, on the other hand, is built on the idea that people process information verbally and visually. It suggests that audio visual information affects learning process positively... Mayer proposed a series of proof-based principles by testing for educational multimedia design"* (Ruiz, et. al., 2009, p.839). Richard E. Mayer's proof-based seven principles on multimedia design containing animation are as follows (Mayer and Moreno, 2002, p. 95-97);

1. The Multimedia Principle: This first principle defines that students experience an in-depth learning through presentations where animation and narration coexist instead of using just words...
2. The Spatial Contiguity Principle: The second principle states that locating image and narration related to that image near each other on presentation provides a better learning experience for students instead of putting images and words far from each other on the presentation screen...
3. The Temporal Contiguity Principle: The third principle describes that students learn better when narration and animation presented synchronously compared to the conditions where they are presented separately...
4. The Coherence Principle: The fourth principle proposes that animation and narrations which exclude extraneous materials such as words, voice (including music) and videos contribute to a better learning experience for students...
5. The Modality Principle: According to the fifth principle, students learn better when animation and verbal narration are presented together instead of presentations containing narration and on-screen text...
6. The Redundancy Principle: The sixth principle puts forward that students experience a thorough learning process where animation and narration are presented instead of a presentation of animation, narration and on-screen text...
7. The Personalization Principle: In this last principle, it is suggested that students learn better when narration is presented in a conversational tone rather than a formal tone while animation and narration take place synchronously...

The number of these principles went from seven to twelve in the second edition of The Cambridge Handbook of Multimedia Learning by Mayer. Besides these original ones, a new set of principles were added. These are signaling, segmenting, pre-training, voice, and image principles. (Mayer, 2009).

One of the new added principles named signaling principle suggests that a better learning process can take place when the important parts of the educational content are emphasized while presenting (p.108). Segmenting principle describes that presenting the content in small segments instead of a whole enhances the learning (p.175). Another principle named pre-training proposes that providing important concepts and names before the lecturing creates a better learning experience (p. 189). Voice principle proposes that using human voice rather than using machine voice in multimedia materials provides better learning (p. 255). Lastly image principle expresses that putting the instructor's image on a corner of the screen during a presentation of a multimedia material doesn't improve learning. On the contrary, this image will create a cognitive load since it's an extraneous visual (p. 258).

Studies on learning have shown that animation is important both for generating motivation and for supporting the storage of information in the long-term memory. Implementing Mayer's Principles of the Cognitive Theory of Multimedia Learning in cartoon movies produced for educational purposes will play a crucial role in transferring the relevant information in a more enduring and successful way. Here in this study, using animation as a tool for knowledge transfer was discussed over visual design and narration in educational cartoon movies and evaluated in terms of Mayer's principles.

Knowledge Transfer in Cartoon Movies

Cartoon movies appear to cooperate with a lot of different disciplines today thanks to their effective and attractive visual narrations, storytelling capabilities and technical features. Sometimes they explain children how human body works, and sometimes they introduce seasons or cultural values to them. Today there are numerous national and international cartoon movies performing educational functions. These cartoon movies with educational content prepared for the needs of different fields have different target groups accordingly. Moving images with educational content prepared for children far from distance education focus and mostly in home environment have been widely produced. In some cartoon movies, the educational content is transferred directly while some of them transfer it disguisedly. During the visual design phase, which is the common ground of these two different content design, it's important to produce cartoon movies in an enjoyable and gripping way without disturbing children's perception (Taş Alicenap and San, 2019, p. 1171). Children learn language development and speech, behavioral styles, social values and, shortly, a life style from the cartoon movies they watch with interest.

TRT Çocuk (Kids) TV channel (TRT Çocuk Official Web Site, n.d.), which began its broadcasting life in Turkey in 2008 and has been broadcasting 24 hours a day only for children, gives a special place to locally produced cartoon movies and series in its broadcast stream. Most of these cartoon series have educational content. "Niloya", "Ege ile Gaga (Ege and Gaga)", "Pepee", "Pırıl", "Elif ve Arkadaşları (Elif and Her Friends)", and "Küçük Hezarfen (Little Hezarfen)" can be listed among those. Each of them provides informative components in different subjects and content. The production which pioneers development of Turkish cartoon series and paves the way for local productions is "Pepee" cartoon series. Bayır and Günşen stated in their article, which is one of the studies on "Pepee" cartoon series, named "*Okul Öncesi Dönem Çocuklarının En Çok İzledikleri Çizgi Filmlerin*

Bilimsel Açılardan Analizi (Scientific Analysis of Cartoons Which Children Watch the Most in Preschool Period)" that during the analysis of cartoon series in terms of scientific information usage, "Pepee" cartoon series took the 1st place among other productions by expressing 72 scientific expressions in 30 episodes (Bayır and Günşen, 2017, p. 751).

"Niloya", which is another production broadcasted on TRT Çocuk channel, tells stories of a girl named Niloya who is around 5 years old and living in a village with her family and her best friend Tospik which is a turtle. This cartoon comprises numerous messages on different subject matters like traditions, social life, family, friendship and love of nature. In a research where the contribution of "Niloya" in value education for preschool children was studied, Karakuş stated that "*totally 157 values were defined in the selected 18 episodes and mostly examples reflecting values such as 'love, sensitivity, tolerance and goodness' were noted and followed by the other values like 'laboriousness, solidarity, respect, responsibility and helpfulness'*" (Karakuş, N., 2015, p. 251).

Another cartoon series broadcasted in the TRT Çocuk channel is named "Pırıl" which is an educatory cartoon series focused on teaching mathematics as well as usage and importance of mathematics in daily life. Pırıl tells the significance and necessity of mathematics through a different subject matter in each episode with the help of her brother and her friends at school. The cartoon series named "Elif ve Arkadaşları" shares what Elif and her friends learn at their kindergarten to which they've just started to attend. NETCO, the production company of the series, states that "almost everything from the activities at kindergarten to the topics that their teacher explains to the children become a content for the cartoon series so that they can become a guide for children, their parents and kindergarten teachers…" (NETCO, n.d.).

The cartoon named "Küçük Hezarfen" is based on the life of Hezarfen Ahmet Çelebi who was a historical figure with an interest in flying. The study made by Selanik and Korkmaz showed that all of the 21 values that take place in the educational program of social studies course are covered in "Küçük Hezarfen" cartoon series; and the most frequently mentioned value is scientificness (Selanik and Korkmaz, 2017, p.55).

As a result, cartoon movies and series are not only entertaining but also educational for children. Cartoon movies present information to children in creative ways by their nature. Information transfer takes place in a disguised and enjoyable manner when cartoon movies' endless creativity meets with children's imaginary world.

"Ege ile Gaga": A Case of Educational Cartoon Series

"Ege ile Gaga" cartoon series tells the adventurous explorations of a boy named Ege and his crow named Gaga in the backyard of their houses. This duo looks for solutions to the problems that they encounter and try to satisfy their curiosity. When they solve the problem, they take notes as their diary entries. The cartoon series is produced by Resimli Filim animation studio (Resimli Filim Animasyon Şirketi, n.d.), which is located at the technopark in Anadolu University (AU) Yunus Emre Campus, and it went on to the finals in Japan Prize Animation Festival under the category of educational cartoon movies. It is aimed to improve children's creativity with the cartoon series "Ege ile Gaga". Here in this study, "Ege ile Gaga" was discussed in terms of visual design and storytelling in educational cartoon series and evaluated in accordance with Mayer's principles.

In every episode of "Ege ile Gaga" independent, short and different themes are focused on. This cartoon series contains two dimensional digitally produced motions. Each episode lasts approximately eight minutes. On the 143rd episode called "Damlaya Damlaya (Drip by Drip)", which was randomly chosen, Ege ile Gaga are concerned about the tomatoes in the garden of their neighbors who aren't at home. They are worried that the tomatoes will dry up; so they create new ideas and solve the problem in the end.

The credits and titles of the cartoon series are prepared by using two dimensional animation technique. In the credits and titles part, Ege and Gaga run after the problems they find out in their backyard. Then these two dimensional images turn into primitive drawings reminding kids' artwork on the pages of the diary on which they make entries after every solution. In the credits and titles part, the images showing main characters, their place of residence and the themes on which cartoon series focus on take place. There is also a music accompanying these images. At the end of these opening credits, the theme of that episode is shown as an on-screen text and a voice-over vocalize these words. For the episode, randomly chosen for this study, the theme of the episode is written as "Damlaya Damlaya (Drip by Drip) as the final scene of opening credits and a voice-over reads those words aloud. Also animation of water drops is seen on these words.

Figure 1.

The title is supported by the visual design. In addition, these words are vocalized synchronously. This usage corresponds to Mayer's pre-training principle. Providing the name of the episode before narration and content gives a prior knowledge about the theme to the audience.

It is seen that an introduction part begins after opening credits and title and in this part, the problem which Ege and Gaga will try to solve is defined and described. This description is made through the dialogues between Ege and Gaga. In the episode chosen for this study, Gaga runs towards Ege who is watering the flowers in the garden and says that "*If tomatoes aren't watered soon, they will dry up. I need your help.*". This sentence describes the problem. Therefore; the audience gets a directive about the theme. However; Gaga flies towards the house instead of the tomatoes while leading Ege and they keep on talking among each other. These elements create extraneous load for the audience.

While moving towards the development phase; a transition scene, on which the name of the cartoon is written, appears and Gaga flies over this text by hitting the last letter. When Gaga hits the last letter on the cartoon name, a bell tone is heard like an alarming sound. This transition is used once again later when the problem solving part of the episode begins. Therefore; the audience receives a directive towards the main theme. This usage corresponds to Mayer's segmenting principle. Ege and Gaga go to the tomato garden and they think that tomatoes will rot if their neighbors won't return soon. At this moment, a notebook page image appears on the screen which shows a children drawing of tomatoes and those tomatoes fall down one by one. By doing this, the audience loses the sight of the place and other distracting images around, and they focus on a screen which clearly depicts the problem. This usage corresponds to Mayer's coherence principle. In this principle Mayer states that "*animation and narrations which exclude extraneous materials such as words, voice (including music) and videos contribute to a better learning experience for students*" (Mayer and Moreno, 2002, p. 95-97).

Figure 2.

A similar notebook page screen is also used while describing their solutions to prevent desiccation of tomatoes (like the scene in which they carry buckets of waters to irrigate tomatoes). However; this solution turn outs to be difficult and tiring. Ege brings out his diary and takes notes of their possible solutions. They think of irrigation by hose. But the hose they have is too short to reach the tomatoe garden. In the meantime Gaga trips over the hose which is laid on the ground among tomatoes and they discover this hose thanks to Gaga's clumsiness. They see holes on this hose and begin to cover them by putting tapes on. But they realize that water trickles away these holes despite the tapes covering them. They see that the trickling water irrigates tomatoes slowly and they remove the tapes immediately. So they puzzle out this system. Then the audience loses sight of the background and other figures, and a scene of a white notebook page without the perception of perspective appears with only depicting the images of irrigation system. This scene is also synchronously narrated by Ege. By this way the working principle of drip irrigation system is told visually with the support of an accompanying verbal narration. This usage corresponds to Mayer's voice principle.

Figure 3.

This usage also matches up with Mayer's multimedia (learning through platforms where both animation and narration take place rather than only narration), temporal contiguity (in depth learning through synchronous verbal narration and animation instead of seperate presentation), coherence (learning through animation and narration which exclude extraneous words (including music), voice and videos), modality (better learning experience by combining animation and verbal narration instead of combining narration and on-screen text) and redundancy principles. Also personalization principle is provided with the conversations of Ege and Gaga by using daily language; and by using a kid's voice representing the character Ege to vocalize the narration on the scenes which depicts notebook page design to describe the subject with simple animations and illustrations. In the conclusion part, learning experience is realized and two dimensional illustration scenes are screened to tell subjects like what drip irrigation is, how it works and what kind of benefits it provides. After these scenes, the ending credits roll with a background image which shows Ege and Gaga eating tomatoes and the episode ends.

As a result, it's seen that the goals of entertaining and educating are met harmoniously in this randomly chosen episode of "Ege ile Gaga". Both the design and the content of the cartoon movie are projected with the correct methods to transfer the objective information. While cinematographic elements, dialogues and spatial images are predominant in the scenes which aims at entertaining; scenes aiming at educating make use of the Cognitive Theory of Multimedia Learning.

It is known that "*using design elements such as warm colors and round shapes triggers positive emotions and they can be used as multimedia learning materials*" (Tien, Chiou and Lee, 2018, p.1). "Ege ile Gaga" cartoon series provide an example of this approach by using round shapes and bright colors in their designs. Also using anthropomorphic characters or designs attracts audience and helps them relate and engage with the motions (Plass, et.al., 2013, p.130). A crow's (Gaga) being the friend with whom Ege solves problems supports this approach as well.

CONCLUSION

There are numerous national and international studies on animation and its usage in the field of education. Here in this study, Mayer's "The Cognitive Theory of Multimedia Learning" and the presentation of educational content in a cartoon by following multimedia design principles were discussed. Preparing educational content and transfer of knowledge requires an interdisciplinary process in which many disciplines cooperates and collaborates, especially in visual communication design.

The humorous and educatory features of animation have gained importance in today's world where digitalization has spread far and wide and education has shifted to distance learning processes. Being accessible for interaction and sharing among different disciplines has made it more attractive. From medicine to mathematics and from fine arts to history, the field of usage for animation in different disciplines has varied in an incredible pace. Therefore; it is believed that the usage of animation in transfer of knowledge and scientific communication will be increasingly discussed in future studies.

REFERENCES

Bayır, E., & Günşen, G. (2017). Okul Öncesi Dönem Çocuklarının En Çok İzledikleri Çizgi Filmlerin Bilimsel Açılardan Analizi. *Trakya Üniversitesi Eğitim Fakültesi Dergisi, 7*(2), 746-761. https://dergipark.org.tr/tr/pub/trkefd/issue/30516/303686

Karakuş, N. (2015). Okul Öncesi Döneme Hitap Eden Tema İçerikli Çizgi Filmlerin Değerler Eğitimine Katkısı Yönünden Değerlendirilmesi (Niloya Örneği). *Değerler Eğitimi Dergisi, 13*(30), 251-277. https://dergipark.org.tr/tr/pub/ded/issue/29166/312334

Mayer, R. E. (2009). Multimedia Learning. New York: Cambridge University Press.

Mayer, R. E., & Moreno, R. (2002, March). Animation as an Aid to Multimedia Learning". *Educational Psychology Review, 14*(1), 87–99. doi:10.1023/A:1013184611077

McClean, P., Johnson, C., Rogers, R., Daniels, L., Reber, J., Slator, B. M., Terpstra, J., & White, A. (2015). Molecular And Cellular Biology Animations: Development And Impact On Student Learning. *Cell Biology Education, 4*(Summer), 169–179. http://citeseerx.ist.psu.edu/viewdoc/download?doi=10.1.1.515.6233&rep=rep1&type=pdf PMID:15917875

NETCO Animasyon Şirketi. (n.d.). Retrieved from https://netco.com.tr/dir/tr/filmler/elif-ve-arkadaslari/

Plass, J. L. (2014). Emotional Design İn Multimedia Learning: Effects Of Shape And Color On Affect And Learning. *Learning and Instruction*, *29*(February), 128–140. doi:10.1016/j.learninstruc.2013.02.006

Resimli Filim Animasyon Şirketi. (n.d.). Retrieved from http://resimlifilim.com/

Ruiz, J. G., Cook, D. A., & Levinson, A. J. (2009). Computer Animations In Medical Education: A Critical Literature Review. Blackwell Publishing Ltd 2. *Medical Education*, *43*(9), 838–846. doi:10.1111/j.1365-2923.2009.03429.x PMID:19709008

Selanik Ay, T., & Korkmaz, Ç. (2017). Sosyal Bilgiler Programında Yer Alan Değerler Ve Kültürel Ögeler Bağlamında "Küçük Hezarfen" Çizgi Filmi. *Batı Anadolu Eğitim Bilimleri Dergisi*, *8*(2), 49-62. https://dergipark.org.tr/tr/pub/baebd/issue/33149/350566

Taş Alicenap, Ç., & San, R. (2019). Çizgi Filmlerin Eğitsel Sunuş Açısından Çoklu Ortam Tasarım İlkeleri ve Sinematizm-Animatizm Kavramları Bağlamında İncelenmesi: The Fixies Çizgi Dizi Örneği. *Gümüşhane Üniversitesi İletişim Fakültesi Elektronik Dergisi*, *7*(2), 1154–1188. doi:10.19145/e-gifder.570937

Tien, L. C., Chiou, C. C., & Lee, Y. S. (2018). Emotional Design in Multimedia Learning: Effects of Multidimensional Concept Maps and Animation on Affect and Learning. *Eurasia Journal of Mathematics, Science and Technology Education*, *14*(12), 1–17. doi:10.29333/ejmste/94229

TRT Cocuk Official Web Site. (n.d.). Retrieved from https://www.trtcocuk.net.tr/

Compilation of References

Adamopoulos, P. (2013). What makes a great MOOC? An interdisciplinary analysis of student retention in online courses. *Proc. of 34th Intern'l Conf. on Info. Systems*. Retrieved from http://pages.stern.nyu.edu/~padamopo/What%20makes%20a%20great%20MOOC.pdf

Adler, M. (2012). The plain language movement. In The Oxford handbook of language and law. Oxford University Press. doi:10.1093/oxfordhb/9780199572120.013.0006

Aesaert, K., van Nijlen, D., Vanderlinde, R., & van Braak, J. (2014). Direct measures of digital information processing and communication skills in primary education: Using item response theory for the development and validation of an ICT competence scale. *Computers & Education, 76*, 168–181. doi:10.1016/j.compedu.2014.03.013

Agora, S. N. S. F. (n.d.). *Agora - where research meets the public.* Retrieved from http://www.snf.ch/en/funding/science-communication/agora/Pages/default.aspx

Akadema. (2019). *Anadolu Üniversitesi Kitlesel Açık Çevrimiçi Ders Platformu*. http://akadema.anadolu.edu.tr

Akbaş, O., & Özdemir, S. M. (2002). Avrupa Birliğinde yaşam boyu öğrenme. *Milli Eğitim Dergisi, 155*(156), 112–126.

Akbulut, K. (2017). Article. *Journal of Financial Solution, 27*, 219-225.

Akca, M., & Aydogan, T. (2015). Online Database Editor Design for Web-Based Distance Learning. *Pamukkale University Journal of Engineering Sciences.*

Akpınar, Y. (2003). Öğretmenlerin yeni bilgi teknolojileri kullanımında Yükseköğretimin etkisi: İstanbul okulları örneği. *The Turkish Online Journal of Educational Technology, 2*(2), 79–96.

Aktan, C. C., & Börü, D. (2007). Kurumsal sosyal sorumluluk. *Kurumsal sosyal sorumluluk: İşletmeler ve sosyal sorumluluk*, 11-36.

Albers, M. J., & Mazur, M. B. (Eds.). (2014). *Content and complexity: information design in technical communication*. Routledge. doi:10.4324/9781410607409

Aldemir, A. (2003). *Bilgiye erişimde yeni yaklaşım: Bilgi Okuryazarlığı*. Aytaç Yıldızeli, Canan Duran, Hatice Kübra Bahşişoğlu.

Allen, L. E., & Taylor, D. M. (2017). The role of the academic library information specialist (LIS) in teaching and learning in the 21st century. *Information Discovery and Delivery*, *45*(1), 1–9. doi:10.1108/IDD-09-2016-0030

Alley, M. (1996). *The craft of scientific writing*. Springer. doi:10.1007/978-1-4757-2482-0

Alomyan, H. (2017). *A Conceptual Framework for Web-Based Learning Design*. International Association for Development of the Information Society.

Altunışık, R., Coşkun, R., Bayraktaroğlu, S. and Yıldırım, E. (2010). *Sosyal Bilimlerde Araştırma Yöntemleri SPSS Uygulamalı* (6. Baskı). Sakarya: Sakarya Yayıncılık.

Apgar, D. (2006). *Risk intelligence: Learning to manage what we don't know*. Harvard Business Press.

Arklan, Ü. (2016). Sosyal medyanın siyasal amaçlı kullanımı: Ağ kuşağının kullanım alışkanlıkları üzerine bir araştırma. *Gümüşhane Üniversitesi İletişim Fakültesi Elektronik Dergisi, 4*(2).

Award, E. R. C. (n.d.). *Public Engagement With Research Award.* Retrieved from https://erc.europa.eu/managing-your-project/public-engagement-research-award

Aydin, A. F. (2017). Sosyal medya ve kurumsal markalaşma ilişkisi: anaakım iletişim kuramları perspektifinden genel bakış. *İletişim Kuram ve Araştırma Dergisi*, (44).

Aydın, C. H. (2011). *Open and Distance Learning Candidates' Perspective*. Pegem Academy.

Baer, K., & Vacarra, J. (2008). *Information design workbook: Graphic approaches, solutions, and inspiration+ 30 case studies*. Rockport Publishers.

Bailey, S. (2014). *Academic writing: A handbook for international students*. Routledge. doi:10.4324/9781315768960

Bakla, A. (2018). Learner-generated materials in a flipped pronunciation class: A sequential explanatory mixed-methods study. *Computers & Education*, *125*, 14–38. doi:10.1016/j.compedu.2018.05.017

Bal, B. (2012). Imaginary Atlases: "Timeless Journeys" with Constant's "New Babylon" and Calvino's "Invisible Cities." *Journal of SKOP, 2.* Retrieved from https://www.e-skop.com/skopdergi/dussel-atlaslar-constantin-new-babylonu-ve-calvinonun-gorunmez-kentleri-ile-zaman-otesi-yolculuklar/581

Balamuralithara, B., & Woods, P. C. (2010). An Investigation on Adoption of the Engineering Simulation Lab Exercise: A Case Study in Multimedia University, *Malaysia. Computer Applications in Engineering Education*, *1*(22), 34–46.

Bates, A. (2005). *Technology, e-Learning and Distance Education*. Routledge. doi:10.4324/9780203463772

Bates, E., Lane, L., & Lange, J. (1993). *Writing clearly: Responding to ESL compositions*. Heinle & Heinle Publishers.

Bayır, E., & Günşen, G. (2017). Okul Öncesi Dönem Çocuklarının En Çok İzledikleri Çizgi Filmlerin Bilimsel Açılardan Analizi. *Trakya Üniversitesi Eğitim Fakültesi Dergisi, 7*(2), 746-761. https://dergipark.org.tr/tr/pub/trkefd/issue/30516/303686

Bayne, S., & Ross, J. (2011). *Digital native'and 'digital immigrant'discourses: a critique. In digital difference*. Brill Sense. doi:10.1007/978-94-6091-580-2_12

Beck, U., Lash, S., & Wynne, B. (1992). *Risk society: Towards a new modernity* (Vol. 17). Sage.

Beghetto, R. A. (2009). Correlates of intellectual risk taking in elementary school science. *Journal of Research in Science Teaching: The Official Journal of the National Association for Research in Science Teaching, 46*(2), 210–223. doi:10.1002/tea.20270

Belanche, D., Casalб, L. V., Orús, C., & Pérez-Rueda, A. (2020). Developing a Learning Network on YouTube: Analysis of Student Satisfaction with a Learner-Generated Content Activity. In *Educational Networking* (pp. 195–231). Springer. doi:10.1007/978-3-030-29973-6_6

Bell, D. (1973). The Coming of Post-Industrial Society: A Venture. In *Social Forecasting*. Harper Colophon Books.

Ben-Chayim, A., & Offir, B. (2019). Model of the Mediating Teacher in Distance Learning Environments: Classes That Combine Asynchronous Distance Learning via Videotaped Lectures. *Journal of Educators Online, 16*(1), n1. doi:10.9743/jeo.2019.16.1.1

Bentley, T. (2003). *Report Writing in Business*. Elsevier.

Berenbaum, M. R. (2017). Communicating about science communication: A brief entomological history. *Annals of the Entomological Society of America, 110*(5), 435–438. doi:10.1093/aesaax060

Bertin, J. (1983). *Semiology of graphics* (W. J. Berg, Trans.). The University of Wisconsin Press.

BFUG. (2003). *Bologna Process between Prague and Berlin, Report to the Ministers of Education of the Signatory Countries*. BFUG.

Biber, D., & Gray, B. (2013). Nominalizing the verb phrase in scientific writing. In B. Aarts, J. Close, G. Leech, & S. Wallis (Eds.), *The verb phrase in English: Investigating recent language change with corpora* (pp. 99–132). Cambridge University. doi:10.1017/CBO9781139060998.006

Bilgici, Z. (2016). *Bilim İletişimi*. Retrieved from https://bilimgenc.tubitak.gov.tr/makale/bilim-iletisimi

Billman, G. E. (2020). Homeostasis: The Underappreciated and Far Too Often Ignored Central Organizing Principle of Physiology. *Frontiers in Physiology*, *11*, 200. doi:10.3389/fphys.2020.00200 PMID:32210840

Blake, J. (2015). Prescriptive-descriptive disjuncture: Rhetorical organisation of research abstracts in information science. In F. Formato & A. Hardie (Eds.), *Conference proceedings of 8th International Corpus linguistics Conference* (pp. 377–8). Lancaster: Lancaster University.

Blake, J. (2018). Corpus-based error detector for Computer Science. In *Proceedings of the Fourth Asia Pacific Corpus Linguistics Conference* (pp.50-54). Takamatsu.

Blake, J. (2020). Genre-specific error detection with multimodal feedback. *RELC Journal, 51*(1), 179-187. https://doi.org/10.1177/0033688219898282

Bonsiepe, G. (1999). Interface: An Approach to Design: In Memoriam Jovita (1941-1998). Jan van Eyck Akademie, Department of Design.

Bordage, G. (2001). Reasons reviewers reject and accept manuscripts: The strengths and weaknesses in medical education reports. *Academic Medicine*, *76*(9), 889–896. doi:10.1097/00001888-200109000-00010 PMID:11553504

Bozkurt, A., Insung, J., Xiao, J., Vladimirschi, V., Schuwer, R., Egorov, G., & Paskevicius, M. (2020, June 5). A global outlook to the interruption of education due to COVID-19 pandemic: Navigating in a time of uncertainty and crisis. *Asian Journal of Distance Education*, 1-126.

Breslow, L., Pritchard, D. E., DeBoer, J., Stump, G. S., Ho, A. D., & Seaton, D. T. (2013). Studying learning in the worldwide classroom: Research into edX's first MOOC. *Research and Practice in Assessment*, *8*, 13–25.

Briggs & Makice. (2012). *Digital Fluency: Building Success in the Digital Age*. Socia Lens.

Britannica. (2020). *Biography of John Ray*. Retrieved 25 September 2020 from https://www.britannica.com/biography/John-Ray-English-naturalist

Brockington, G., & Mesquita, L. (2016). The consequences of bad science communication. *Revista da Biologia*, *15*(1), 29–34. doi:10.7594/revbio.15.01.03

Browner, W. S. (2012). *Publishing and Presenting Clinical Research* (3rd ed.). Lippincott Williams & Wilkins.

Brown, H. (2000). *Principles of Language Learning and Teaching*. Prentice Hall.

Browning, L. D., Satre, A. S., Stephens, K. K., & Sornes, J. O. (2004). *Information & communication technologies in action: Linking theory & narratives of practice*. Copenhagen Business School Press.

Brown, T. (2009). *Change by Design: How Design Thinking Transforms Organizations and Inspires Innovation*. HarperCollins Publishers.

Brush, T., & Saye, J. (2014). An Instructional Model to Support Problem-Based Historical Inquiry: *The Persistent Issues in History Network. The Interdisciplinary Journal of Problem-Based Learning, 8*(1), 3–7. doi:10.7771/1541-5015.1409

Bucchi, M., & Saracino, B. (2016). "Visual Science Literacy" Images and Public Understanding of Science in the Digital Age. *Science Communication, 38*(6), 812–819. doi:10.1177/1075547016677833

Buchanan, R. (1992). *Wicked Problems in Design Thinking*. doi:10.2307/1511637

Budak, Y. (2009). Yaşamboyu öğrenme ve ilköğretim programlarının hedeflemesi gereken insan tipi. *Gazi Üniversitesi Gazi Eğitim Fakültesi Dergisi, 29*(3).

Burns, M. (2020). *eLearning Industry*. We Are All Distance Learners: Online Education Across The Globe—Part 2: https://elearningindustry.com/distance-learners-online-education-across-the-globe-part-2

Burt, M., & Kiparsky, C. (1978). Global and local mistakes. In J. Schumann & N. Stenson (Eds.), *New Frontiers in Second Language Learning* (pp. 71–79). Newbury House.

Cairo, A. (2012). *The Functional Art: An introduction to information graphics and visualization*. New Riders.

Cakmak, A. C. (2013). Evaluation of Distance Learning Service by Students: An Application at Karabuk University. *Istanbul Commerce University Journal of Social Sciences, 12*(23), 263-287.

Çalık, D., & Çınar, Ö. P. (2009). Geçmişten günümüze bilgi yaklaşımları bilgi toplumu ve internet. In XIV. Türkiye'de İnternet Konferansı (pp.77-88). Bilgi Üniversitesi.

Canagarajah, A. S. (1996). "Nondiscursive" requirements in academic publishing, material resources of periphery scholars, and the politics of knowledge production. *Written Communication, 13*(4), 435–472. doi:10.1177/0741088396013004001

Carlson, J. R., & Zmud, R. W. (1999). Channel expansion theory & the experiential nature of media richness perceptions. *Academy of Management Journal, 42*, 153–170.

Cassin, R. (2002). La formation Tout au Long de la Vie, Séminaire La formation professionnelle. Ecole Nationale d'Administration Paris Comission Européenne. Union Europeenne. Education et Formation Tout au Longue de la Vie. Education et Formation Jeunesse.

Castells, M. (2005). *Information Age: Economy, Society and Culture - Rise of Network Society* (E. Kilic, Trans.). İstanbul Bilgi University Publications.

Çebi, A., & Ve Reisoğlu, İ. (2019). Öğretmen adaylarının dijital yeterliklerinin geliştirilmesine yönelik bir eğitim etkinliği: Böte ve diğer branşlardaki öğretmen adaylarının görüşleri. *Eğitim Teknolojisi Kuram ve Uygulama, 9*(2), 539–565.

CEDEFOP. (2001). *Memorandum on Lifelong Learning Consultation Process: A Review of Member State and EEA Country Reports*. CEDEFOP.

CERI. (2001). Centre for Educational Research and Innovation. Education Policy Analysis. OECD.

Cernikova, M., Smahel, D., & Ve Wright, M. F. (2015). The impact of digital media on health: Children's perspectives. *International Journal of Public Health*, (60), 131–137. PMID:25601331

Champaign, J., Colvin, K. F., Liu, A., Fredericks, C., Seaton, D., & Pritchard, D. E. (2014). Correlating skill and improvement in 2 MOOCs with a student's time on tasks. *Proc. of First ACM Conf. on Lrng. @ Scale Conf*, 11-20. 10.1145/2556325.2566250

Chen, X., Barnett, D. R., & Stephens, C. (2013). Fad or future: The advantages and challenges of massive open online courses (MOOCs). Presented at Research to Practice Conf. in Adult and Higher Ed, Lindenwood University.

Cheng, F. F., Chiu, C. C., Wu, C. S., & Tsaih, D. C. (2017). The influence of learning style on satisfaction and learning effectiveness in the asynchronous web-based learning system. *Library Hi Tech*, *35*(4), 473–489. doi:10.1108/LHT-12-2016-0151

Chigona, A. (2018). Digital fluency: Necessary competence for teaching and learning in connected classrooms. *The African. Journal of Information Systems*, *10*(4).

Christiansen, E. T., Kuure, L., Mørch, A., & Lindström, B. (2013). Problem Based Learning for The 21st Century New Practices and Learning Environments. Aalborg University Press.

Cidral, W. A., Oliveira, T., Di Felice, M., & Aparicio, M. (2018). E-learning success determinants: Brazilian empirical study. *Computers & Education*, *122*, 273–290. doi:10.1016/j.compedu.2017.12.001

Conferences, P. C. S. T. (n.d.). *History of PCST conferences.* Retrieved from https://www.pcst.co/conferences

COST. (n.d.). *About COST.* Retrieved from https://www.cost.eu/who-we-are/about-cost/

Costa, A., & Kallick, B. (2007). *Describing 16 habits of mind.* Retrieved November 13,2020, from http://schoolnet.org.za/teach10/resources/eo/resources/thinking/Habits_of_Mind.pdf

Costa, C., Sousa, C., Rogado, J., & Henriques, S. (2017). Playing digital security: Youth voices on their digital rights. *International Journal of Game-Based Learning*, *7*(3), 11–25. doi:10.4018/IJGBL.2017070102

Council, B. D. (2015). *What is the framework for innovation?...* Design Council's evolved Double Diamond: https://www.designcouncil.org.uk/news-opinion/what-framework-innovation-design-councils-evolved-double-diamond

Coutinho, C. (2007). Infusing technology in pre-service teacher education programs in portugal: A study with weblogs. In *Proceedings of the 18th International Conference of the Society for Information Technology & Teacher Education.* Chesapeake, VA: AACE.

Cozma, A. (2019). *Sidlaurea*. Why is Storytelling at the core of Design Thinking?: https://sidlaurea.com/2019/09/22/why-storytelling-is-at-the-core-of-design-thinking/

Creswell, J. W. (2007). *Qualitative inquiry and research method: Choosing among five approaches*. Academic Press.

Creswell, J. W. (2005). *Educational Research: Planning, Conducting, and Evaluating Qualitative and Quantitative Research* (2nd ed.). Pearson Education.

Creswell, J. W. (2013). Research design: Qualitative, quantitative, and mixed methods approaches. *Sage (Atlanta, Ga.)*.

Crismond, D. P., & Adams, R. S. (2012). The informed design teaching and learning matrix. *Journal of Engineering Education, 101*(4), 738–797. doi:10.1002/j.2168-9830.2012.tb01127.x

Çubukçu, Z., Tosuntaş, Ş. B., & İnci, T. (2018). Eğitim programları aracılığıyla dijital yeterliklerin kazandırılması Dijital Yeterlikler Çerçevesi örneği. Presented at the *International Symposium on Multidisciplinary Academic Studies*, Berlin.

da Silva, J. L., de Oliveira, R. F., Bertin, L. D., de Oliveira, P. D., Pires, F. B., Spadaccini, A. F., ... & de Almeida Pires-Oliveira, D. A. (2018). Physical inactivity and associated factors in adults of a basic health unit: An observational cross-sectional study. *Manual Therapy, Posturology & Rehabilitation Journal, 16*.

Dahlstrom, M. F. (2014). Using Narratives And Storytelling To Communicate Science With Nonexpert Audiences. *Proceedings of the National Academy of Sciences of the United States of America, 111*(Supplement_4), 13614–13620. Advance online publication. doi:10.1073/pnas.1320645111 PMID:25225368

Dale, C., & Povey, G. (2009). An evaluation of learner-generated content and podcasting. *Journal of Hospitality, Leisure, Sport and Tourism Education, 8*(1), 117–123. doi:10.3794/johlste.81.214

Dave, R. H. (1976). *Foundtions of Lifelong Education*. Pergamon Press.

Davies, S. R., & Horst, M. (2016). *Science Communication: Culture, Identity and Citizenship*. Palgrave Macmillan. doi:10.1057/978-1-137-50366-4

de Oliveira, L. L., & Bocchini, M. O. (2013). *Associação entre design da informação e linguagem simples em materiais informativos das prefeituras de São Paulo (1990), Campinas (2000) e Várzea Paulista (2009)*. Retrieved November 16, 2016, from http://pdf.blucher.com.br.s3-sa-east-1.amazonaws.com/designproceedings/cidi/CIDI-4.pdf

Dede, C. (2005). Planning for neomillennial learning styles. *EDUCAUSE Quarterly, 28*(1), 7–12.

Demir, K. (2018). Öğretmen adaylarının dijital akıcılıklarının incelenmesi. Doktora tezi, Anadolu Üniversitesi Eğitim Bilimleri Enstitüsü, Eskişehir.

DePrince, A. E. III, & Hammond, J. R. (2011). Coupled cluster theory on graphics processing units I. The coupled cluster doubles method. *Journal of Chemical Theory and Computation, 7*(5), 1287–1295. doi:10.1021/ct100584w PMID:26610123

Di Bitetti, M. S., & Ferreras, J. A. (2017). Publish (in English) or perish: The effect on citation rate of using languages other than English in scientific publications. *Ambio, 46*(1), 121–127. doi:10.100713280-016-0820-7 PMID:27686730

Dinsmore, D. L., & Alexander, P. A. (2012). A critical discussion of deep and surface processing: What it means, how it is measured, the role of context, and model specification. *Educational Psychology Review, 24*(4), 499–567. doi:10.100710648-012-9198-7

Dodd, W. D. (2015). *Teaching and Learning in the Digital Age.* https://webcpm.com/articles/2015/09/01/teaching-learning-technologies.aspx

Dönmez, G. (2019). *Lise Öğrencilerinin Bilgi Güvenliği Farkındalığı İle Dijital Okuryazarlığı Arasındaki İlişkinin İncelenmesi.* Yüksek Lisans Tezi. Hacettepe Üniversitesi Eğitim Bilimleri Enstitüsü.

Doyle, C. S. (1994). *Information literacy in an information society: A concept for the information age.* Syracuse University.

d-school. (2020). *Product Discovery Methods.* Interview for Empathy: https://pdmethods.com/interview-for-empathy/

Duffy, P. (2008). Engaging the YouTube Google-eyed generation: Strategies for using Web 2.0 in teaching and learning. *Electronic Journal of E-learning, 6*(2), 119–130.

Duly, H., & Burt, M. (1974). Natural sequences in child second language acquisition. *Language Learning, 24*, 23–40.

Duman, T. (2001). Avrupa Birliği Eğitim Programları ‘Leonardo da Vinci Programı’. *Milli Eğitim Dergisi*, (150).

Dur, B. I. U. (2014). Data visualization and infographics in visual communication design education at the age of information. *Journal of Arts and Humanities, 3*(5), 39–50.

Duupdates. (2020, June 24). *Duupdates.* UNESCO report 2020: Online education implemented during covid-19 are not inclusive: https://www.duupdates.in/unesco-report-2020-online-education-implemented-during-covid-19-are-not-inclusive/

Dynel, M. (2009). *Humorous garden-paths: A pragmatic-cognitive study.* Cambridge Scholars Publishing.

Edge, J. (1990). *Mistakes and correction.* Longman.

Ekmekçioğlu, D., & Özkaraman, M. (2017). Defining Self-Organizing Systems as a Design Problem. *Online Journal of Art and Design, 5*(4), 89-101. Retrieved from http://www.adjournal.net/articles/54/546.pdf

Elliott, A. B. (1983). *Errors in English*. National University of Singapore Press.

Entwistle, N., & Peterson, E. (2004). Conceptions of learning and knowledge in higher education: Relationships with study behavior and influences of learning environments. *International Journal of Educational Research, 41*(6), 407–428. doi:10.1016/j.ijer.2005.08.009

ERC Science Communication. (n.d.). *#Science Communication*. Retrieved from https://erc.europa.eu/tags/science-communication

Erdoğan, İ. (1997). *Communication, Domination, Introduction to Struggle*. Imge Publishing.

Erdoğan, İ. (2005). *Understanding Communication*. Erk Publications.

Erkan, H. (1998). *Bilgi Toplumu ve Ekonomik Gelişme*. Türkiye İş Bankası Kültür Yayınları.

ESCI. (n.d.). *European Science Communication Institute – Our Services*. Retrieved from https://www.esci.eu/service/

ESOF. (n.d.). *Past ESOF*. Retrieved from https://www.esof.eu/en/about/esof-in-trieste.html

Estrada, F. C. R., & Davis, L. S. (2015). Improving visual communication of science through the incorporation of graphic design theories and practices into science communication. *Science Communication, 37*(1), 140–148. doi:10.1177/1075547014562914

European Commission. (1994). *White Paper on Growth, Competitiveness, and Employment: The Challenges and Ways Forward into the 21st Century*, Retrieved from https://www.cvce.eu/content/publication/1997/10/13/b0633a76-4cd7-497f-9da1-4db3dbbb56e8/publishable_en.pdf

European Commission. (1995). *European Commission, Teaching and Learning: Towards the Learning Society*. Retrieved from https://europa.eu/documents/comm/white_papers/pdf/com95_590_en.pdf

European Commission. (2002). *European Commission, European Report on Quality of Lifelong Learning, Brussels*. Retrieved from http://www.aic.lv/bolona/Bologna/contrib/EU/report_qual%20LLL.pdf

European Commission. (n.d.). *Horizon 2020*. Retrieved from https://ec.europa.eu/programmes/horizon2020/what-horizon-2020

European Commission-EC. (2006). The Key Competences for Lifelong Learning – A European Framework. Official Journal of the European Union, 394(1).

European Commission-EC. (2007). Key competences for lifelong learning - A European framework. Luxembourg: Author.

European Research Council. (n.d.). *Mission*. Retrieved from https://erc.europa.eu/about-erc/mission

European Science Communication Institute. (n.d.). *European Science Communication Institute - About Us*. Retrieved from https://www.esci.eu/about-us/

Evans, D. (2015). *Risk intelligence: How to live with uncertainty*. Simon and Schuster.

Evliyaoğlu, G. (1987). *Communication Psychology, Psychological Communication*. Journalists Association Publications.

Faber, J. (2017). Writing scientific manuscripts: Most common mistakes. *Dental Press Journal of Orthodontics*, *22*(5), 113–117. doi:10.1590/2177-6709.22.5.113-117.sar PMID:29160351

Fearing, B. E., & Sparrow, W. K. (Eds.). (1989). *Technical writing: Theory and practice*. Modern Language Association.

Ferrari, A. (2012). Digital competence in practice: An analysis of frameworks. Sevilla, Spain: Academic Press.

Ferrari, A. (2013). *DIGICOMP: A framework for developing and understanding digital competence in Europe*. Luxembourg: JRC Scientiðc and Policy Reports EUR26036EN.

Ferris, D. R. (2006). Does error feedback help student writers? New evidence on the short-and long-term effects of written error correction. In K. Hyland & F. Hyland (Eds.), *Feedback in second language writing: Contexts and issues* (pp. 81–104)., doi:10.1017/CBO9781139524742.007

Ferris, D. R. (2011). *Treatment of error in second language student writing*. The University of Michigan Press. doi:10.3998/mpub.2173290

Fetter, L. C. (2011). *Revistas, design editorial e retórica tipográfica: a experiência da revista Trip* (Unpublished doctoral dissertation). Retrieved November 20, 2019, from https://www.lume.ufrgs.br/handle/10183/30193

Fidalgo-Neto, AA, Tornaghi, AJC, Meirelles, RMS, Berçot, FF, & Xavier, LL, Castro, MFA ve Alves, LA. (2009). Brezilya ilk ve ortaokullarında bilgisayar kullanımı. *Bilgisayar ve Eğitim*, *53*(3), 677–685.

Findeli, A. (2001). *Rethinking Design Education for the 21st Century: Theoretical*. Methodological, and Ethical Discussion. Retrieved from https://www.mitpressjournals.org/doi/pdf/10.1162/07479360152103796

Fırat, M., Yurdakul, I. K. & Ersoy, A. (2014). Bir eğitim teknolojisi araştırmasına dayalı olarak karma yöntem araştırması deneyimi. *Eğitimde Nitel Araştırmalar Dergisi*, *2*(1).

Fischer, G., & Konomi, S. (2005). Innovative media in support of distributed intelligence and lifelong learning. In *Proceedings of the Third IEEE International Workshop on Wireless and Mobile Technologies in Education*, (pp. 3-10). Los Alamitos, CA: IEEE Computer Society. 10.1109/WMTE.2005.35

Fischer, H. (2018). *Clareza em textos de e-gov: uma questão de cidadania*. Com Clareza.

Fischhoff, B., & Scheufele, D. A. (2013). The science of science communication. *Proceedings of the National Academy of Sciences of the United States of America*, *110*(Supplement 3), 14031–14032. doi:10.1073/pnas.1312080110 PMID:23942127

Fiske, J. (2003). İletişim Çalışmalarına Giriş. Ankara: Science and Art.

Flowerdew, J. (2014). *Academic Discourse*. Routledge. doi:10.4324/9781315838069

Foundation, I. D. (2020). *Focus Groups*. What are Focus Groups: https://www.interaction-design.org/literature/topics/focus-groups

Franco, J. R. (2017). O Design de Informação e a narrativa diagramática em Whale Hunt de Jonathan Harris. *METAgraphias, 2*(3).

Franco, J. R., & Borges, P. M. (2019). Diagrammatic relations of probative strength and inferential progression through semiotics. *Semiotica, 2019*(228), 77–89. doi:10.1515em-2018-0088

Frascara, J. (2004). *Communication design: principles, methods, and practice*. Allworth Communications, Inc.

FREREF. (2004). *Lifelong Learning The Challenges*. Printed by l'Imprimerie Ecully Graphic.

Fries, C. (1945). *Teaching and Learning English as a Foreign Language*. University of Michigan.

Fulk, J., Steinfield, C. W., Schmitz, J., & Power, J. G. (1987). A Social Information Processing Model of Media Use in Organizations. *Communication Research, 14*(5), 529–552. doi:10.1177/009365087014005005

Funston, F., & Wagner, S. (2010). *Surviving and thriving in uncertainty: Creating the risk intelligent enterprise*. John Wiley & Sons.

FWF WissKomm. (n.d.). *Science Communication Programme (WissKomm)*. Retrieved from https://www.fwf.ac.at/en/research-funding/fwf-programmes/science-communication-programme-wisskomm/

Gallardo-Echenique, E. E., de Oliveira, J. M., Marqués-Molias, L., Esteve-Mon, F., Wang, Y., & Baker, R. (2015). Digital competence in the knowledge society. *Journal of Online Learning and Teaching, 11*(1).

Gecis, Z. (2019). *UX Collective*. The 2x2 matrix: how to be more systematic about the decisions you make: https://uxdesign.cc/2-x-2-decision-matrix-bf17364a4993

Gefen, R. (1979). The analysis of pupil's errors. *English Teachers'. Journal, 22*, 16–24.

Gibbons, S. (2016). *Nielsen Norman Group*. Design Thinking 101: https://www.nngroup.com/articles/design-thinking/

Gigerenzer, G. (2015). *Risk savvy: How to make good decisions*. Penguin.

Göksan, T. S., Uzundurukan, S., & Keskin, S. N. (2009). Yaşam boyu öğrenme ve avrupa birliği'nin yaşam boyu öğrenme programlari. *İnşaat Mühendisliği Eğitimi Sempozyumu, 143*.

Göktas, Y., Yildirim, Z., & Yildirim, S. (2008). A Review of ICT Related Courses in Pre-service Teacher Education Programs. *Asia Pacific Education Review, 9*(2), 168–179. doi:10.1007/BF03026497

Gordon, J., Halász, G., Krawczyk, M., Leney, T., Michel, A., Pepper, D., ... & Wiśniewski, J. (2009). Key competences in Europe: Opening doors for lifelong learners across the school curriculum and teacher education. *CASE Network Reports*, (87).

Gracıous, A., & Shyla, A. J. (2012). Multıple ıntellıgence and dıgıtal learnıng awareness of prospectıve teachers. *Turkish Online Journal of Distance Education-TOJDE*, *11*(3), 112–118.

Grainger, B. (2013). *Massive Open Online Course (MOOC) Report*. University of London Intern'l. Academy. Retrieved from http://www.londoninternational.ac.uk/sites/default/files/documents/mooc_report-2013.pdf

Graves, H., & Graves, R. (2012). *A Strategic Guide to Technical Communication*. Broadview.

Groff, J. (2013). *Technology-rich innovative learning environments*. Paris, France: OECD Publishing. Retrieved November 23, 2020, from http://www.oecd.org/edu/ceri/TechnologyRich%20Innovative%20Learning%20Environments%20by%20Jennifer%20Groff.pdf

Gross, A. G., Harmon, J. E., Reidy, M., & Reidy, M. S. (2002). *Communicating science: The scientific article from the 17th century to the present*. Oxford University Press on Demand.

Guasch, T., Alvarez, I., & Espasa, A. (2010). University teacher competencies in a virtual teaching / learning environment: Analysis of a teacher training experience. *Teaching and Teacher Education*, *26*(2), 199–206. doi:10.1016/j.tate.2009.02.018

Güleç, İ., & Çelik, S., & Demirhan, B. (2012). Yaşam Boyu Öğrenme Nedir? Kavram ve Kapsami Üzerine Bir Değerlendirme. *Sakarya University Journal of Education*, *2*(3), 34–48.

Gulnar, B. (2008). Design, Development and Evaluation Stages of Computer and Internet Supported Distant Learning Programs (Sample of Suzep)]. *Selcuk University Journal of the Institute of Social Sciences*, (19), 259–271.

Gündoğan, N. (2003). Avrupa Birliği'ne Üye Ülkelerde Bir İstihdam Politikası Aracı Olarak "Yaşamboyu Öğrenme" ve Bazı Örnek Program ve Uygulamalar. *Kamu İş, 7*(2).

Hall, G. M. (2011). *How to Write a Paper* (5th ed.). Wiley – Blackwell.

Halliday, M. A. K. (1994). *An Introduction to Functional Grammar* (2nd ed.). Edward Arnold.

Halliday, M. A. K., & Martin, J. R. (1993). *Writing science: Literacy and discursive power*. The Falmer Press.

Hanauer, D. A., Mei, Q., Vydiswaran, V. V., Singh, K., Landis-Lewis, Z., & Weng, C. (2019). Complexities, variations, and errors of numbering within clinical notes: The potential impact on information extraction and cohort-identification. *BMC Medical Informatics and Decision Making*, *19*(3), 75. doi:10.118612911-019-0784-1 PMID:30944012

Harris, J., Mishra, P., & Koehler, M. (2009). Teachers' technological pedagogical content knowledge and learning activity types: Curriculum-based technology integration reframed. *Journal of Research on Technology in Education*, *41*(4), 393–416. doi:10.1080/15391523.2009.10782536

Heinich, R., Molenda, M., Russell, J., & Smaldino, S. (1999). *Instructional Media and Technologies for Learning* (6th ed.). Prentice-Hall.

Hendricks, B. (2020). *Study.com*. Bodystorming in Design Thinking: Definition, Purpose & Example: https://study.com/academy/lesson/bodystorming-in-design-thinking-definition-purpose-example.html

Hilgartner, S. (1990). The Dominant View of Popularization: Conceptual Problems, Political Uses. *Social Studies of Science*, *20*(3), 519–539. doi:10.1177/030631290020003006

Hinds, P., & Kiesler, S. (1995). Communication across boundaries: Work, structure, & use of communication technologies in a large organization. *Organization Science*, *6*(4), 373–393. doi:10.1287/orsc.6.4.373

Hmelo-Silver, C. E. (2004). Problem-Based Learning: What and How Do Students Learn? *Educational Psychology Review*, *16*(3), 235–266. doi:10.1023/B:EDPR.0000034022.16470.f3

Hodges, C., Moore, S., Lockee, B., Trust, T., & Bond, A. (2020, March 27). *Educausereview*. Retrieved from The Difference Between Emergency Remote Teaching and Online Learning: https://er.educause.edu/articles/2020/3/the-difference-between-emergency-remote-teaching-and-online-learning#:~:text=In%20contrast%20to%20experiences%20that,mode%20due%20to%20crisis%20circumstances

Holmes, A. (2002). *Lifelong learning*. Capstone Publishing.

Holtz, M. (2009). Nominalization in scientific discourse: A corpus-based study of abstracts and research articles. In *Proceedings of the 5th Corpus Linguistics Conference, #341*. Available at: http://ucrel.lancs.ac.uk/publications/cl2009/

Horizon 2020 Work Programme 2014 - 2015. (n.d.). *16. Science with and for Society Revised.* Retrieved from https://ec.europa.eu/research/participants/data/ref/h2020/wp/2014_2015/main/h2020-wp1415-swfs_en.pdf

Horizon 2020 Work Programme 2016 - 2017. (n.d.). *16. Science with and for Society.* Retrieved from https://ec.europa.eu/research/participants/data/ref/h2020/wp/2016_2017/main/h2020-wp1617-swfs_en.pdf

Horizon 2020 Work Programme 2018 - 2020. (n.d.). *16. Science with and for Society.* Retrieved from https://ec.europa.eu/research/participants/data/ref/h2020/wp/2018-2020/main/h2020-wp1820-swfs_en.pdf

Horizon 2020. (n.d.). *Science with and for Society.* Retrieved from https://ec.europa.eu/programmes/horizon2020/en/h2020-section/science-and-society

Horn, R. E. (2000). Information design: Emergence of a new profession. In R. Jacobson (Ed.), *Information design*. MIT Press.

Howe, N., & Strauss, W. (2000). *Millennials rising: The next great generation*. Vintage Original.

Hsi, S. (2007). Conceptualizing learning from the everyday activities of digital kids. *International Journal of Science Education*, *29*(12), 1509–1529. doi:10.1080/09500690701494076

Hsu, L. (2017). Enhancing college students' satisfaction and learning interest when the instructor uses a web-based platform while teaching. *American Journal of Educational Research*, *5*(1), 18–24.

Hulcr, J., Dharampal, P. S., Hamm, R. L., Pearson, G., & Gibson, C. (2019). Influence Is Power: Strategic Communication for Entomologists. *American Entomologist (Lanham, Md.)*, *65*(2), 92–96. doi:10.1093/ae/tmz024

Hyland, K. (2002a). Authority and invisibility: Authorial identity in academic writing. *Journal of Pragmatics*, *34*(8), 1091–1112. doi:10.1016/S0378-2166(02)00035-8

Hyland, K. (2002b). Options of identity in academic writing. *ELT Journal*, *56*(4), 351–358. doi:10.1093/elt/56.4.351

Hyland, K. (2006). *English for Academic Purposes: An Advanced Resource Book*. Routledge. doi:10.4324/9780203006603

Idrus, Z., Abidin, Z. Z. S., & Hashım, R. & Omar, N. (2010). Awareness in networked collaborative environment: a comparative study on the usage of digital elements. *Proceedings of the 9th WSEAS International Conference on applıcatıons of computer engineering*, 236-241.

Illingworth, S., & Allen, G. (2016). *Effective Science Communication*. Bristol, UK: IOP Publishing. Retrieved from http://webcache.googleusercontent.com/search?q=cache:reKcQk72mzkJ:library.iyte.edu.tr/dosya/kitap/efective.pdf+&cd=1&hl=en&ct=clnk&gl=tr

Ilomäki, L., Paavola, S., Lakkala, M., & Kantosalo, A. (2016). Digital competence–an emergent boundary concept for policy and educational research. *Education and Information Technologies*, *21*(3), 655–679. doi:10.100710639-014-9346-4

Ingels, M. B. (2006). *Legal English communication skills*. Acco.

Insee. (2020). *Insee*. Statistics and studies: https://www.insee.fr/en/statistiques

İpşiroğlu, Z. (1997). *Eğitimde yeni arayışlar*. Adam Yayınları.

Isman, A. (2011). *Distance Education*. Pegem Akademi.

Jackson, L., Jones, S., & Rodriguez, R. C. (2010). Faculty actions that result in student satisfaction in online courses. *Journal of Asynchronous Learning Networks*, *14*(4), 78–96. doi:10.24059/olj.v14i4.129

Jackson, S. (2006). Learning to live: The relationship between lifelong learning and lifelong illness. *International Journal of Lifelong Education*, *25*(1), 51–73. doi:10.1080/02601370500309501

James, C. (1998). *Errors in language learning and use*. Longman.

Janssen, J., Stoyanov, S., Ferrari, A., Punie, Y., Pannekeet, K., & Sloep, P. (2013). Experts' views on digital competence: Commonalities and differences. *Computers & Education, 68*, 473–481. doi:10.1016/j.compedu.2013.06.008

Jarvis, P. (2004). *Adult Education and Lifelong Learning: Theory and Practice* (3rd ed.). Routledge. doi:10.4324/9780203561560

Jensen, J. L., McDaniel, M. A., Woodard, S. M., & Kummer, T. A. (2014). Teaching to the tes or testing to teach: Exams requiring higher order thinking skills encourage greater conceptual understanding. *Educational Psychology Review, 26*(2), 307–329. doi:10.100710648-013-9248-9

Jonas, W. (2001). A Scenario for Design. *Design Issues, 17*(2), 64-84. Retrieved from https://www.mitpressjournals.org/doi/pdf/10.1162/07479360152383796

Jones, C., Ramanau, R., Cross, S., & Healing, G. (2010). Net generation or digital natives: Is there a distinct new generation entering university? *Computers & Education, 54*(3), 722–732. doi:10.1016/j.compedu.2009.09.022

Kaercher, A. L. (2015). *Adaptação automática de interface de acordo com características do usuário.* UniRitter. Retrieved December 2, 2019, from https://bit.ly/35zrL8K

Kant, I. (2019). *Answering the question: What is enlightenment?* Strelbytskyy Multimedia Publishing.

Karakas, F., Manisaligil, A., & Sarigollu, E. (2015). Management learning at the speed of life: Designing reflective, creative, and collaborative spaces for millenials. *International Journal of Management Education, 13*(3), 237–248. doi:10.1016/j.ijme.2015.07.001

Karakuş, N. (2015). Okul Öncesi Döneme Hitap Eden Tema İçerikli Çizgi Filmlerin Değerler Eğitimine Katkısı Yönünden Değerlendirilmesi (Niloya Örneği). *Değerler Eğitimi Dergisi, 13*(30), 251-277. https://dergipark.org.tr/tr/pub/ded/issue/29166/312334

Kaysi, F., & Aydemir, E. (2017). Evaluation of interaction dimensions in distance learning processes. *Journal of Social Sciences, 4*(11), 778–790. doi:10.16990/SOBIDER.3442

Kellert, S. R. (1993). Values and perceptions of invertebrates. *Conservation Biology, 7*(4), 845–855. doi:10.1046/j.1523-1739.1993.740845.x

Kellogg, R. T. (2008). Training Writing Skills: A Cognitive Developmental Perspective. *Journal of Writing Research, 1*(1), 1-26. Retrieved from http://citeseerx.ist.psu.edu/viewdoc/download?doi=10.1.1.569.8025&rep=rep1&type=pdf

Khoury, C. K., Kisel, Y., Kantar, M., Barber, E., Ricciardi, V., Klirs, C., & Valiño, Á. (2019). Science–graphic art partnerships to increase research impact. *Communications Biology, 2*(1), 1–5. doi:10.103842003-019-0516-1 PMID:31396575

Kilic, F. (2014). Awareness and Cognitive Load Levels of Teacher Candidates towards Student Products Made by Digital Storytelling. *Turkish Online Journal of Distance Education, 15*(3), 94–107. doi:10.17718/tojde.90227

Kılınç, H. (2016). *İçerik Üreten Öğrenenlerin Özelliklerinin Değerlendirilmesi. Yüksek Lisans Tezi* (Master Thesis). Anadolu Üniversitesi, Sosyal Bilimler Enstitüsü, Eskişehir.

Kingdom, W. (2014). English Grammar and Style: Good writing Practice. *Medical Writing, 23*(1), 55–57. doi:10.1179/2047480613Z.000000000172

King, N. (2004). Using templates in the thematic analysis of text. In C. Cassell & G. Symon (Eds.), *Essential guide to qualitative methods in organizational research* (pp. 256–270). Sage. doi:10.4135/9781446280119.n21

King, R. C., & Xia, W. (1999). Media appropriateness: Effects of experience on communication media choice. In K. E. Kendall (Ed.), *Emerging information technologies: Improving decision, cooperation, & infrastructure* (pp. 143–175). Sage. doi:10.4135/9781483345505.n7

Kinnunen, T., Leisma, H., Machunik, M., Kakkonen, T., & Lebrun, J.-L. (2012). SWAN - Scientific Writing AssistaNt. A Tool for Helping Scholars to Write Reader-Friendly Manuscripts. In *Proceedings of the 13th Conference of the European Chapter of the Association for Computational Linguistics*, (pp. 20–24). Stroudsburg, PA: Association for Computational Linguistics.

Kip, B. (2007). *Çevrimiçi Öğenenlerin Farklı Destek Ortamlarını Kullanma Sıklıklarıyla Sosyal Bulunuşluk Algıları Arasındaki İlişki* (Unpublished master's thesis). Anadolu University, Eskişehir.

Klamma, R., Cao, Y., & Spaniol, M. (2007). *Watching the Blogosphere: Knowledge sharing in Web 2.0.* Paper presented at the International Conference on Weblogs and Social Media, Boulder, CO. https://www.icwsm.org/papers/2--Klamma-Cao-Spaniol.pdf

Klein, D., Koltin, O., Peleg, M., Portnoy, I., & Greenbaum, D. (2017). Science and Law Separated by Impenetrable Language Barriers: Overcoming Impediments to Much Needed Interactions. *AJOB Neuroscience, 8*(1), 37–39. doi:10.1080/21507740.2017.1285831

Konte, R. (2019). *Anatomy of a Bite. Project Description.* Retrieved December 3, 2019, from https://meetings.ami.org/2017/project/5-2/

Korkmaz, O., Cakir, R., & Tan, S. S. (2015). The Effect of Students' Readiness for E-Learning and Satisfaction Levels on Academic Success]. *Journal of Kirsehir Education Faculty, 16*(3).

Kozikoğlu, İ., & Altunova, N. (2018). Öğretmen Adaylarının 21. Yüzyıl Becerilerine İlişkin Öz-yeterlik Algılarının Yaşam Boyu Öğrenme Eğilimlerini Yordama Gücü. *Journal of Higher Education & Science/Yüksekögretim ve Bilim Dergisi, 12*(3).

Krašna, M., Bratina, T., & Bedrac, B. (2011). Digital competences in education-digital security. In *INTED2011 Proceedings* (pp. 1634-1641). IATED.

Krause, K. (2017). A framework for visual communication at Nature. *Public Understanding of Science (Bristol, England)*, *26*(1), 15–24. doi:10.1177/0963662516640966 PMID:27117485

Kulich, J. (1982). Lifelong education and the universities: A Canadian perspective. *International Journal of Lifelong Education*, *1*(2), 123–142. doi:10.1080/0260137820010204

Kuş, G. (2018). Doğal Afetler Sonrasinda Insanlarin Ilkyardim Bilgi Ve Beceri Düzeyi: Ne Yapabiliriz? *Anadolu University of Sciences & Technology-B: Theoretical Sciences*, *6*(2).

Laan, K. V. (2012). *The Insider's Guide to Technical Writing*. XML Press.

Lado, R. (1957). *Linguistics across Cultures*. University of Michigan.

Lambert, C., Philp, J., & Nakamura, S. (2017). Learner-generated content and engagement in second language task performance. *Language Teaching Research*, *21*(6), 665–680. doi:10.1177/1362168816683559

Lambert, C., & Zhang, G. (2019). Engagement in the Use of English and Chinese as Foreign Languages: The Role of Learner-Generated Content in Instructional Task Design. *Modern Language Journal*, *103*(2), 391–411. doi:10.1111/modl.12560

Lamport, L. (1994). *LATEX: a document preparation system: user's guide and reference manual*. Addison-wesley.

Lave, J., & Wenger, E. (1991). *Situated learning: Legitimate peripheral participation*. Cambridge University Press. doi:10.1017/CBO9780511815355

Lee, M. J. W, McLoughlin, C. & Chan, A. (2008). Talk the talk: Learner-generated podcasts as catalysts for knowledge creation. *British Journal of Educational Technology*, *39*(3), 501–521. DOİ: doi:10.1111/j.1467-8535.2007.00746.x

Lee, K. (2018). Everyone already has their community beyond the screen: Reconceptualizing online learning and expanding boundaries. *Educational Technology Research and Development*, *66*(5), 1255–1268. doi:10.100711423-018-9613-y

Lee, M. J. W., & McLoughlin, C. (2007). Teaching and Learning in the Web 2.0 Era: Empowering Students through Learner-Generated Content. *International Journal of Instructional Technology and Distance Learning*, *4*(10).

Lee, M., & Friedrich, T. (2011). Continuously reaffirmed, subtly accommodated, obviously missing and fallaciously critiqued: Ideologies in UNESCO's lifelong learning policy. *International Journal of Lifelong Education*, *30*(2), 151–169. doi:10.1080/02601370.2010.547619

Lee, S. J., & Reeves, T. C. (2007). Edgar Dale: A significant contributor to the field of educational technology. *Educational Technology*, *47*(6), 56.

Lengrand, P. (1989). Lifelong education: growth of the concept. In *Lifelong Education for Adults* (pp. 5–9). Pergamon. doi:10.1016/B978-0-08-030851-7.50008-4

Lenhart, A., & Madden, M. (2005). *Teen content creators and consumers.* Pew Internet & American Life Project.

Lennon, M., Kirsch, I., Von Davier, M., Wagner, M., & Yamamoto, K. (2003). *Feasibility Study for the PISA ICT Literacy Assessment. Report to Network A* Educational Testing Service.

Leonard, A. A., Elizabeth, F., & Marta, K. (2007). V-Learning Distance Education in the 21st Century Through 3D Virtual Learning Environments. Academic Press.

Liedtka, J. (2014). Innovative ways companies are using. *Strategy and Leadership*, *42*(2), 40–45. doi:10.1108/SL-01-2014-0004

Lipton, R. (2011). *The practical guide to information design.* John Wiley & Sons.

Liu, A., Li, Y., & Ye, H. (2018). Digital fluency and social media use: An empirical ınvestigation of WeChat use. *Proceedings of the 51st Hawaii International Conference on System Sciences*, 503-512. 10.24251/HICSS.2018.065

Lohr, L. L. (2007). *Creating graphics for learning and performance: Lessons in visual literacy.* Prentice Hall Press.

Loveland, T., & Dunn, D. (2014, May/June). Teaching engineering habits of mind in technology education. *Technology and Engineering Teacher*, *73*(8), 13–19.

Lucia, S., Navarro, D., Karg, B., Sarnago, H., & Lucia, O. (2021). Deep Learning-Based Model Predictive Control for Resonant Power Converters. *IEEE Transactions on Industrial Informatics*, *17*(1), 409–420. doi:10.1109/TII.2020.2969729

Luckin, R. (2006). Understanding learning contexts as ecologies of resources: From the zone of proximal development to learner generated contexts. *World Conference on E-Learning in Corporate, Government, Healthcare, and Higher Education, 2006*(1), 2195-2202.

Luckin, R., du Boulay, B., Smith, H., Underwood, J., Fitzpatrick, G., Holmberg, J., Kerawalla, L., Tunley, H., Brewster, D., & Pearce, D. (2005). Using Mobile Technology to Create Flexible Learning Contexts. *Journal of Interactive Media in Education*, *2005*(2), 21. doi:10.5334/2005-22

Lundby, K., & Ronning, H. (2002). Media, Culture, Communication: Interpretation of Modernity Through Media Culture. In Media, Culture, Politics. Ankara: Alp Press.

MacDougall, M., Riley, S. C., Cameron, H. S., & McKinstry, B. (2008). Halos and Horns in the Assessment of Undergraduate Medical Students: A Consistency-Based Approach. *Journal of Applied Quantitative Methods*, *3*(2), 116–128.

Maiorana, F. A., & Mayer, H. F. (2018). How to avoid common errors in writing scientific manuscripts. *European Journal of Plastic Surgery*, *41*(5), 489–494. doi:10.100700238-018-1418-z

Markel, M. (2012). *Technical Communication* (10th ed.). Bedford/St. Martins.

Marra, R., Jonassen, D., Palmer, B., & Luft, S. (2014). Why Problem-Based Learning Works: Theoretical Foundations. *Journal on Excellence in College Teaching*, *25*, 221–238.

Matthews, J. R., & Matthews, W. (2007). *Successful scientific writing: A step-by-step guide for the biological and medical sciences* (3rd ed.). Cambridge University Press. doi:10.1017/CBO9780511816185

Mayer, R. E. (2003). *The Promise of Multimedia Learning: Using the Same Instructional Design Methods Across Different Media, Learning and Instruction.* Retrieved from https://www.journals.elsevier.com/learning-and-instruction

Mayer, R. E. (2009). Multimedia Learning. New York: Cambridge University Press.

Mayer, R. E., & Moreno, R. (2002, March). Animation as an Aid to Multimedia Learning". *Educational Psychology Review, 14*(1), 87–99. doi:10.1023/A:1013184611077

Mazur, B. (2000). Revisiting plain language. *Technical Communication (Washington), 47*(2), 205–211.

McClean, P., Johnson, C., Rogers, R., Daniels, L., Reber, J., Slator, B. M., Terpstra, J., & White, A. (2015). Molecular And Cellular Biology Animations: Development And Impact On Student Learning. *Cell Biology Education, 4*(Summer), 169–179. http://citeseerx.ist.psu.edu/viewdoc/download?doi=10.1.1.515.6233&rep=rep1&type=pdf PMID:15917875

McIntyreL. (2018). *Post-truth.* MIT Press.

McKercher, B., Law, R., Weber, K., Song, H., & Hsu, C. (2007). Why referees reject manuscripts. *Journal of Hospitality & Tourism Research (Washington, D.C.), 31*(4), 455–470. doi:10.1177/1096348007302355

McKinnon, M., & Vos, J. (2016). Crossing A Threshold. In M. C. A. van der Sanden & M. J. de Vries (Eds.), Science and Technology Education and Communication: Seeking Energy (pp. 27-45). Sense Publishers.

Merrill, M. D. (2012). First Principles of Instruction: Identifying and Designing Effective, Efficient, and Engaging Instruction. Hoboken, NJ: Pfeiffer (John Wiley & Sons).

Metin, A. E., Karaman, A., & Sastim, Y. A. (n.d.). Student Perspectives on Distance Education and the Assessment of the Efficiency of Distance Education English Courses: Banaz Vocational College. *Karabuk University Journal of Social Sciences Institute, 7*(2), 640-652.

Miller, C., & Bartlett, J. (2012). 'Digital fluency': Towards young people's critical use of the internet. *Journal of Information Literacy, 6*(2), 35–55. doi:10.11645/6.2.1714

Miller, L. C. (2010). *Make me A STORY.* Stenhouse Publishers. Retrieved from https://books.google.com.tr/books?hl=tr&lr=&id=dIINvn2RGUAC&oi=fnd&pg=PT8&dq=miller+2010+writing&ots=LIIQ6IGhLu&sig=bfQEpsNyIZ59AQLsFr_6SbEX1yE&redir_esc=y#v=onepage&q=miller%202010%20writing&f=false

Milli Eğitim Bakanlığı (MEB). (2018). *Öğretim Programları.* http://mufredat.meb.gov.tr/Programlar.aspx

Milli Eğitim Bakanlığı (MEB). (2020). http://hayatboyu.meb.gov.tr/hayat-boyu-ogrenme/

Minard, C. J. (1861). *Tableaux graphiques et cartes figuratives, Bibliothèque numérique patrimoniale des ponts et chaussées*. Retrieved December 13, 2019, from https://patrimoine.enpc.fr/document/ENPC01_Fol_10975

Mishra, P., Koehler, M. J., & Henriksen, D. (2011). The seven trans-disciplinary habits of mind: Extending the TPACK framework towards 21st century learning. *Educational Technology*, 22–28.

Moore, M., & Kearsley, G. (2005). *Distance education: a systems view*. Wadsworth.

Morrison, J., Roth McDuffie, A., & French, B. (2015, May). Identifying key components of teaching and learning in a STEM school. *School Science and Mathematics*, *115*(5), 244–255. doi:10.1111sm.12126

Mutlu, M. E., Özöğüt Erorta, Ö., Kip Kayabaş, B., & Kayabaş, İ. (2014). Anadolu Üniversitesi Açıköğretim Sisteminde e-Öğrenmenin Gelişimi. In Açıköğretimle 30 Yıl (pp. 1-58). Anadolu University Publications.

Narayan, V. (2011). *Learner-generated content as a pedagogical change agent*. Ascilite.

National Research Council, Committee on Information Technology Literacy. (1999). *Being fluent with information technology*. National Academies Press.

Nesselhauf, N. (2003). The Use of Collocations by Advanced Learners of English and Some Implications for Teaching. *Applied Linguistics*, *24*(2), 223–242. doi:10.1093/applin/24.2.223

NETCO Animasyon Şirketi. (n.d.). Retrieved from https://netco.com.tr/dir/tr/filmler/elif-ve-arkadaslari/

Network for the Public Communication of Science and Technology. (n.d.). *About the PCST Network*. Retrieved from https://www.pcst.co/about

Network. (n.d.). In *Merriam-Webster dictionary online*. Retrieved from https://www.merriam-webster.com/dictionary/network

Networking. (n.d.). In *Cambridge Dictionary online*. Retrieved from https://dictionary.cambridge.org/dictionary/english/networking

Newman, J., & Oh, S. (2014, June 13). 8 things you should know about MOOCs. *Chron. of Higher Ed.* Retrieved from http://chronicle.com/article/8-Things-You-Should-Know- About/146901/

Newman, M. (2003). *A pilot systematic review and meta-analysis on the effectiveness of problem-based learning*. Retrieved November 10, 2020 from https://www.heacademy.ac.uk/assets/documents/subjects/medev/pbl_report.pdf

News, U. (2020). *Global perspective*. Startling disparities in digital learning emerge as COVID-19 spreads: UN education agency: https://news.un.org/en/story/2020/04/1062232

Nichols, M. (2016). Reading and studying on the screen: An overview of literature towards good learning design practice. *Journal of Open, Flexible, and Distance Learning*, *20*(1), 33–43.

Niessen, S. (2013). *What is digital fluency*. http://www.shuananiessen.ca/what-is-digital-fluency/

NSERC Canada. (n.d.). *Science Communication Skills grant (pilot)*. Retrieved from https://www.nserc-crsng.gc.ca/Promoter-Promotion/ScienceComm_eng.asp

O'Connor, M. (2002). *Writing Successfully in Science*. E. & F.M. Spon. doi:10.4324/9780203478684

O'Reilly, T. (2005). *What is Web 2.0: Design patterns and business models for the next generation of software*. https://www.oreilly.com/pub/a/Web2/archive/what-is-Web-20.html

Ober, S. (2007). *Contemporary Business Communication* (7th ed.). Houghton Mifflin.

Oblinger, D., Oblinger, J. L., & Lippincott, J. K. (2005). Educating the net generation. EDUCAUSE.

Odabaşı, H. F. (2000). Akademik eskimişlik ve sürekli mesleki gelişim. *Anadolu Üniversitesi Eğitim Fakültesi Dergisi*, *10*(2), 161–166.

OECD. (2004). *Lifelong Learning, Policy Brief*. Retrieved from http://www.oecd.org/dataoecd/17/11/29478789.pdf

OECD. (2013). *PISA results: Ready to learn: Students' engagement, drive and Self-beliefs* (Vol. 3). OECD Publishing. doi:10.1787/9789264201170-

Ohanian, S. (2008). In a Time of Universal Deceit, Telling the Truth Is a Revolutionary Act. *What is Authentic Educational Reform? Pushing Against the Compassionate Conservative Agenda*, 199.

Oktem, N. (1995). *State and Legal Philosophical Movements with the Data of Sociology and Philosophy*. Der Published.

Olive, T. (2012). Working Memory in Writing. Ed. Virginia Wise Berninger. In *Past, Present and Future Contributions of Cognitive Writing Research to Cognitive Psychology*. New York: Psychology Press Taylor & Francis Group. Retrieved from https://books.google.com.tr/books?id=MoaTPgbuAccC&pg=PA485&lpg=PA485&dq=olive+2004+writing&source=bl&ots=uYZgIiAI60&sig=okR1hmIcFcTh3pKIknS6KgcCdG8&hl=tr&sa=X&ved=0ahUKEwjZ1pPJaTYAhWRIOwKHZELBvIQ6AEITjAF#v=onepage&q=olive%202004%20writing&f=false

Orr, T., & Yamazaki, A. K. (2004). Twenty problems frequently found in English research papers authored by Japanese researchers. In *International Professional Communication Conference Proceedings* (pp. 23–35). 10.1109/IPCC.2004.1375270

Özdemir Çakır, H. (2009). *Kurumsal sosyal sorumluluğun marka imajına etkisi*. Academic Press.

Ozerbaş, M.A. (2012). Öğretimi Ayrıntılama Teorisi, Elaboration Theory. *GEFAD/GUJGEF, 32*(2).

Özmusul, M. (2018). Eğitim Ve Yetiştirme 2020 Stratejisi: Hedef Göstergelere Göre Ülkelerin Durumu. *Journal of European Education*, *2*(1).

Ozonur, M., & Tekdal, M. (2004). *The Effect of Web Based Asynchronous Distance Education Application Designed Based on the Detailed Theory of Teaching on Academic Achievement of University Students*. XIII. National Educational Sciences Congress, Inonu Universty, Malatya, Turkey.

Öztürk, Y. E., & Kıraç, R. (n.d.). Sağlık ve Hastalık. *Scientific Developments*, 382.

Parlak, B. (2017). Dijital çağda eğitim: Olanaklar ve uygulamalar üzerine bir analiz. *Süleyman Demirel Üniversitesi İktisadi ve İdari Bilimler Fakültesi Dergisi*, (22), 1741–1759.

Paul, R. & Elder, L. (2006). *Minik eleştirel düşünme kılavuzu: Kavramlar ve araçlar* (M. Bektaş, Çev). Academic Press.

PCST Network. (n.d.). *The PCST Network Constitution*. Retrieved from https://www.pcst.co/doc/PCST.Constitution.2016-04-27.pdf

Peat, J., Elliott, E., Baur, L., & Keena, V. (2002). *Scientific Writing: Easy When You Know How*. BMJ Books. doi:10.1002/9781118708019

Peirce, C. S. (2012). *Charles S. Peirce, Selected Writings*. Courier Corporation.

Pérez-Mateo, M., Maina, M. F., Romero, M., & Guitert, M. (2011). Learner Generated Content: quality from students' point of view. *World Conference on Educational Multimedia, Hypermedia and Telecommunications*, 2011(1), 2520-2529.

Petelin, R. (2010). Considering plain language: Issues and initiatives. *Corporate Communications*, *15*(2), 205–216. doi:10.1108/13563281011037964

Pettersson, R. (2010). Information design–principles and guidelines. *Journal of Visual Literacy*, *29*(2), 167–182. doi:10.1080/23796529.2010.11674679

Pierson, D. J. (2004). The top 10 reasons why manuscripts are not accepted for publication. *Respiratory Care*, *49*(10), 1246–1252. PMID:15447812

Pinho, I. D. C., & Lima, M. S. (2013). Teacher's digital fluency: A new competence for foreigh language teaching. *Revista Brasileira de Lingüística Aplicada*, *13*(3), 711–739. doi:10.1590/S1984-63982013005000014

Plain Language Association International (PLAIN). (n.d.). *What is Plain Language?* Retrieved December 10, 2019, from https://plainlanguagenetwork.org/plain-language/what-is-plain-language/

Plass, J. L. (2014). Emotional Design İn Multimedia Learning: Effects Of Shape And Color On Affect And Learning. *Learning and Instruction*, *29*(February), 128–140. doi:10.1016/j.learninstruc.2013.02.006

Polino, C., & Castelfranchi, Y. (2012). The 'Communicative Turn' in Contemporary Techno-science: Latin American Approaches and Global Tendencies. In B. Schiele, M. Claessens, & S. Shi (Eds.), *Science Communication in the World: Practices, Theories and Trends* (pp. 3–17). Springer. doi:10.1007/978-94-007-4279-6_1

Potter, M. C., Wyble, B., Hagmann, C. E., & McCourt, E. S. (2014). Detecting meaning in RSVP at 13 ms per picture. *Attention, Perception & Psychophysics*, *76*(2), 270–279. doi:10.375813414-013-0605-z PMID:24374558

Prensky, M. (2001). Digital Natives, Digital Immigrants. From On the Horizon, 9(5).

Prettyman, S. S., Ward, C. L., Jauk, D., & Awad, G. (2012, Fall). 21st century learners: Voices of students in a one-to-one STEM environment. *Journal of Applied Learning Technology*, *2*(4), 6–15.

Putnam, A. R. (2011). Problem-Based Teaching and Learning in Technology Education. *5th Slovakian-Hungarian Joint Symposium on Applied Machine Intelligence and Informatics.*

QMPP Project. (2009). *QMPP Handbook Quality Management of Peer Production.* Author.

QTurkey Bilim İletişimi. (n.d.). *Science Communication.* Retrieved from https://www.qturkey.org/science-communication

QTurkey. (n.d.). *About QTurkey Community.* Retrieved from https://www.qturkey.org/about

Rajamanickam, V. (2005). Infographics: Handout. National Institute of Design, 1-14.

Rea, P. & Irving, D. (2004). *Cinema and Short Films on Video- 1, Pre- Production* (S. Taylaner, Trans.). İstanbul: Es.

Reid, W. A., Duvall, E., & Evans, P. (2005). Can we influence medical students' approaches to learning? *Medical Teacher*, *27*(5), 401–407. doi:10.1080/01421590500136410 PMID:16147792

Reigeluth, C. M. (1998). *Elaboration Theory: Scope and Sequence Decisions for Quality Instruction.* Retrieved from https://www.reigeluth.net/pubsinsttheor

Reiser, R., & Gagne, R. (1983). Selecting Media for Instruction. Academic Press.

Resimli Filim Animasyon Şirketi. (n.d.). Retrieved from http://resimlifilim.com/

Richards, J. C., & Schmidt, R. (2002). *Longman's Language Teaching & Applied Linguistics* (3rd ed.). Pearson Education Limited.

Richardson, J. C., & Newby, T. (2006). The Role of Students' Cognitive Engagement in Online Learning. *American Journal of Distance Education*, *20*(1), 23–37. doi:10.120715389286ajde2001_3

Rodriguez, J. A. O., Ladd, M., & Brandis, D. (2020). Abdominal Thrust Maneuver. In *StatPearls*. StatPearls Publishing.

Rosen, K. (2020, November 12). *attentionfwd.* 2020: Lost year of schooling or year one in a new era of education? https://attentionfwd.com/insights/2020-lost-year-of-schooling-or-year-one-in-a-new-era-of-education/

Ruiz, J. G., Cook, D. A., & Levinson, A. J. (2009). Computer Animations In Medical Education: A Critical Literature Review. Blackwell Publishing Ltd 2. *Medical Education*, *43*(9), 838–846. doi:10.1111/j.1365-2923.2009.03429.x PMID:19709008

Sageev, P., & Romanowski, C. J. (2001). A Message from Recent Engineering Graduates in the Workplace: Results of a Survey on Technical Communication Skills. *Journal of Engineering Education*, *90*(4), 685–693. doi:10.1002/j.2168-9830.2001.tb00660.x

Sağlam, M., Özüdoğru, O. F., & Çiray, F. (2011). Avrupa Birliği eğitim politikaları ve Türk Eğitim Sistemi'ne etkileri. *Yüzüncü Yıl Üniversitesi Eğitim Fakültesi Dergisi*, *8*(1), 87–109.

Sahin, T. Y., & Yildirim, S. (1999). Educational Technologies and Material Development. Academic Press.

Sáiz-Manzanares, M. C. (2018). *E-Project Based Learning en Terapia Ocupacional: una Aplication en la Asignatura "Estimulacion Temprana*. Servicio de Publicaciones de la Universidad de Burgos.

Santaella, L., Franco, J. D. O. R., De Bona, C. M., & Espínola, V. (2016). Por uma semiótica não-tergiversante: Análise do site Conductor-MTA. me. *LÍBERO*, (28), 37–76.

Savery, J. (2006). Overview of Problem-based Learning: Devinition and Distinction Interdisciplinary. *Journal Problem-Based Learning*, *1*(1), 9–20. doi:10.7771/1541-5015.1002

Scerri, E. R. (1998). The evolution of the periodic system. *Scientific American*, *279*(3), 78–83. doi:10.1038cientificamerican0998-78

Schein, M., Farndon, J. R., & Fingerhut, A. (2000). Why should a surgeon publish? *British Journal of Surgery*, *87*(1), 3–5. doi:10.1046/j.1365-2168.2000.01373.x PMID:10606903

Schirrmeister, R. T., Springenberg, J. T., Fiederer, L. D. J., Glasstetter, M., Eggensperger, K., Tangermann, M., Hutter, F., Burgard, W., & Ball, T. (2017). Deep learning with convolutional neural networks for EEG decoding and visualization. *Human Brain Mapping*, *38*(11), 5391–5420. doi:10.1002/hbm.23730 PMID:28782865

Science Communication Network. (n.d.). *Science Communication Network: Who We Are, What We Do*. Retrieved from http://sciencecommunicationnetwork.org/

Science Communication Work Group. (n.d.). *Science Communication Work Group Presentation*. Retrieved from https://www.euroscience.org/structure/science-communication-workgroup/

Scott, J. (2011). Social network analysis: Developments, advances, and prospects. *Social Network Analysis and Mining*, *1*(1), 21–26. doi:10.100713278-010-0012-6

Seaton, D. T., Bergner, Y., Chuang, I., Mitros, P., & Pritchard, D. E. (2014). Who does what in a massive open online course? *Communications of the ACM*, *57*(4), 58–65. doi:10.1145/2500876

Sekret, I. V., & Kommers, P. A. (2014). Conceptual issues of the digital competence development in the framework of the council of the European Union. *Conference proceedings of 10th International Scientific Conference on Distance Learning in Applied Informatics DIVAI2014*, 183-195.

Selaník Ay, T., & Korkmaz, Ç. (2017). Sosyal Bilgiler Programında Yer Alan Değerler Ve Kültürel Ögeler Bağlamında "Küçük Hezarfen" Çizgi Filmi. *Batı Anadolu Eğitim Bilimleri Dergisi, 8*(2), 49-62. https://dergipark.org.tr/tr/pub/baebd/issue/33149/350566

Selinker, L. (1972). Interlanguage. *International Review of Applied Linguistics in Language Teaching, 10*(3), 209–231.

Sener, J. (2007). In search of student-generated content in online education. *E-Mentor, 21*, 1-8.

Siagian, M. V., Saragih, S., & Sinaga, B. (2019). Development of Learning Materials Oriented on Problem-Based Learning Model to Improve Students' Mathematical Problem Solving Ability and Metacognition Ability. *International Electronic Journal of Mathematics Education, 14*(2), 331–340. doi:10.29333/iejme/5717

Siddiq, F. (2018, October). *A Comparison Between Digital Competence in Two Nordic Countries' National Curricula and an International Framework: Inspecting their Readiness for 21st Century Education*. Academic Press.

Siebert, D. (2010). *Eğitimde Dijital Devrim.* Deuthsche Welle.

Siemens, G. (2004). Connectivism: A learning theory for the digital age. International. *Journal of Instructional Technology and Distance Learning, 2*(1), 3–10.

Siemens, G. (2010). *Teaching in Social and Technological Networks*. Connectivism.

Simionescu, M., & Simion, E. (2004). Scientific Lingua Franca and National Languages at the Crossroads. In P. Drenth & J. Schroots (Eds.), *ALLEA Biennial Yearbook. Critical Topics in Science and Scholarship* (pp. 129–133). ALLEA. Available at https://allea.org/wp-content/uploads/2016/02/Simionescu_-Lingua_Franca.pdf

Sishley, S. (2020, May). *E-Learning Solutions*. Emergency Remote Learning Compared to Online Learning: https://learningsolutionsmag.com/articles/emergency-remote-learning-compared-to-online-learning

Skrypnyk, O., Joksimović, S., Kovanović, V., Gašević, D., & Dawson, S. (2015). Roles of course facilitators, learners, and technology in the flow of information of a cMOOC. *The International Review of Research in Open and Distributed Learning, 16*(3). Advance online publication. doi:10.19173/irrodl.v16i3.2170

SNSF Science Communication. (n.d.). *Science Communication.* Retrieved from http://www.snf.ch/en/funding/science-communication/Pages/default.aspx

SNSF. (n.d.). *Mission.* Retrieved from http://www.snf.ch/en/theSNSF/organisation/Pages/default.aspx

Song, Y., Lung, K. M., Ogata, H., Yang, Y., Hasnine, N., & Mouri, K. (2019). Bridging in-class and real life vocabulary learning using a mobile learner-generated content tool: A case study. In *Teacher forum proceedings of the 23rd Global Chinese Conference on Computers in Education (GCCCE2019)* (pp. 176-186). Central China Normal University.

Sparrow, J. (2018). *Digital fluency: Preparing students to create big, bold problems.* https://er.educause.edu/articles/2018/3/digital-fluency-preparing-students-to-create-big-bold-problems

Spencer, K. (2015). *What is Digital Fluency*? [blog post]. http://blog.core-ed.org/blog/2015/10/what-is-digital-fluency.html

Stegmaier, A. (2013). Species Hunt. *National Geographic Magazine.* Retrieved December 13, 2019, from https://www.nationalgeographic.com/magazine/2013/04/species-hunt/

Stevens, E. (2020, November 18). *Careerfoundry.* What Is Design Thinking? A Comprehensive Beginner's Guide: https://careerfoundry.com/en/blog/ux-design/what-is-design-thinking-everything-you-need-to-know-to-get-started/

Subiela-Hernández, B.-J. (2017). Diseño de la información y progreso social: Prospectiva, perspectiva y retos. *El Profesional de la Información, 26*(6), 1019–1024. doi:10.3145/epi.2017.nov.01

Sujana, A., Permanasari, W., & Sopandi, A. M. (2012). Article. *Jurnal Pendidikan IPA Indonesia, 1*(1), 91–97. Retrieved November 24, 2020, from: http://journal.unnes.ac.id/nju/index.php/jpii%0ALITERASI

Sürmelioğlu, Y., & Seferoğlu, S. S. (2019). Yükseköğretim Öğrencilerinin Dijital Ayakizi Farkındalık ve Yaşantılarının İncelenmesi. *World Journal on educational technology. Current Issues, 11*(1), 48–64.

Swales, J. M. (1997). English as *Tyrannosaurus rex. World Englishes, 16*(3), 373–382. doi:10.1111/1467-971X.00071

Tamer, M. G. (2014). Yaşam boyu öğrenme için ulusal yeterlilikler çerçevesi: Avrupa ve Türkiye örneği. *International Journal of Curriculum and Instructional Studies, 3*(5).

Tan, C. L., & Morris, J. S. (2006). Undergraduate college students, laptop computers, and lifelong learning. *The Journal of General Education, 54*(4), 316–338. doi:10.1353/jge.2006.0013

Tang, R., & John, S. (1999). The 'I' in identity: Exploring writer identity in student academic writing through first person pronoun. *English for Specific Purposes, 18*, 23–39. doi:10.1016/S0889-4906(99)00009-5

Tapscott, D. (1998). *Growing up digital: The rise of the net generation.* McGraw Hill.

Taş Alícenap, Ç., & San, R. (2019). Çizgi Filmlerin Eğitsel Sunuş Açısından Çoklu Ortam Tasarım İlkeleri ve Sinematizm-Animatizm Kavramları Bağlamında İncelenmesi: The Fixies Çizgi Dizi Örneği. *Gümüşhane Üniversitesi İletişim Fakültesi Elektronik Dergisi, 7*(2), 1154–1188. doi:10.19145/e-gifder.570937

Taylor, C. (1979). What's wrong with negative liberty? *The Idea of Freedom,182.*

Tekerek, M. ve Tekerek, A. (2013). A research on students' information security awareness. *Turkish Journal of Education, 2*(3), 61–70.

Tezci, E., & Gürol, A. (2003). Constructivist Instructional Design and Creativity. *TOJET, 2*(1), 50-55. Retrieved from http://www.tojet.net/articles/v2i1/218.pdf

The Euroscience Open Forum. (n.d.). *ESOF in a Nutshell*. Retrieved from https://www.esof.eu/en/about-esof/aims.html

Thomas, D., & Brown, J. S. (2011). *A new culture of learning: Cultivating the imagination for a world of constant change. Lexington*. CreateSpace.

Thrower, P. (2012). *Eight reasons I rejected your article*. Elsevier online. Available at: http://www.elsevier.com/connect/8-reasons-i-rejected-your-article

Tien, L. C., Chiou, C. C., & Lee, Y. S. (2018). Emotional Design in Multimedia Learning: Effects of Multidimensional Concept Maps and Animation on Affect and Learning. *Eurasia Journal of Mathematics, Science and Technology Education, 14*(12), 1–17. doi:10.29333/ejmste/94229

Til, U. D. A. (2020). Yeni Koronavirüs hastalığı Hakkında bilinmesi gerekenler. *Ayrıntı Dergisi, 8*(85).

Today, E. (2019). *Why Science Communication Ought to Be Part of the Job for Entomologists*. Retrieved December 13, 2019, from https://entomologytoday.org/2017/02/01/why-science-communication-ought-to-be-part-of-the-job-for-entomologists/

Toffler, A. (1996). Third Wave (3rd ed.). Istanbul: Altin Kitaplar.

Topper, A., & Lancaster, S. (2013). Common challenges and experiences of school districts that are implementing one-to-one computing initiatives. *Computers in the Schools, 30*(4), 346–358. doi:10.1080/07380569.2013.844640

Toprak, M., & Erdogan, A. (2012). Yaşamboyu Öğrenme: Kavram, Politika, Araçlar ve Uygulama [Lifelong Learning: Concept, Policy, Instruments and Implementation]. *Journal of Higher Education and Science, 2*(2), 69. doi:10.5961/jhes.2012.036

Toprak, M., & Erdoğan, A. (2012). Yaşamboyu öğrenme: Kavram, politika, araçlar ve uygulama. *Yükseköğretim ve Bilim, 2*(2), 69–91.

Torres, R. M., & Cano, J. (1995). Learning style: A factor to critical thinking? *Journal of Agricultural Education, 36*(4), 55–62. doi:10.5032/jae.1995.04055

Touchie, H. Y. (1986). Second language learning errors: Their types, causes, and treatment. *JALT Journal, 8*(1), 75-80.

Traphagan, T., Kucsera, J. V., & Kishi, K. (2010). Impact of class lecture webcasting on attendance and learning. *Educational Technology Research and Development, 58*(1), 19–37. doi:10.100711423-009-9128-7

Tribull, C. M. (2017). Sequential science: A guide to communication through comics. *Annals of the Entomological Society of America, 110*(5), 457–466. doi:10.1093/aesaax046

TRT Cocuk Official Web Site. (n.d.). Retrieved from https://www.trtcocuk.net.tr/

Tschimmel, K., Santos, J., Loyens, D., Jacinto, A., Monteiro, R., & Valença, M. (2018). *D-Think: Design Thinking Applied to Education and Training*. European Commission.

TUBITAK Bilim Genc. (n.d.). *Bilim Genc*. Retrieved from https://bilimgenc.tubitak.gov.tr/dergimiz/hakkimizda-0

TUBITAK. (n.d.). *Who We Are?* Retrieved from https://www.tubitak.gov.tr/en/about-us/content-who-we-are

Tufte, E. R. (2007). *The visual display of quantitative information*. Graphics Press.

Tufte, E. R., & Robins, D. (1997). *Visual explanations*. Graphics Press.

Türnüklü, A. (2000). *Eğitimbilim Araştırmalarında Etkin Olarak Kullanılabilecek Nitel Bir Araştırma Tekniği: Görüşme. Kuram ve Uygulamada Eğitim Yönetimi Dergisi. 24*. PegemA Yayıncılık.

UN (United Nations). (2015a). *Transforming our world: The 2030 Agenda for sustainable development*. New York: United Nations. Retrieved 25 February 2019 from https://sustainabledevelopment.un.org/ post2015/transformingourworld

UNESCO. (2020). *COVID-19 Impact on Education*. Education: From disruption to recovery: https://en.unesco.org/covid19/educationresponse

Ural, Y. Ç. (2017). Solunum, kalp ve dolaşım sistemiyle ilgili acil durumlar ve ilk yardım. *Pegem Atıf İndeksi*, 110-142.

Urhan, V. (2011). *Pedigree of M. Foucault and Knowledge / Power Relationship*. Retrieved from http://www.uludag.edu.tr/~kaygi/dergi/articles/9_99-118.pdf

Usta, İ., Uğur, S., Okur, M. R., & Kuş, G. (2019). A Study on The Use of Social Media of Open Education System Students: The Example of AofAnadolum. *Kastamonu Education Journal, 27*(2).

Vail, R. (2013). *Who is afraid of a big bad MOOC?* Colorado Mesa University. Retrieved from http://cmubiznews.org/uploads/facultyinsights/Vail_Who_is_afraid_of_a_MOOC_Febru ary_2013.pdf

Van Merrienboer, J. J. G., & Kirschner, P. A. (2013). *Ten steps to complex learning (2nd Reviewed.)*. Routledge.

van Weijen, D. (2014). *How to overcome common obstacles to publishing in English*. Elsevier online. Available at: https://www.elsevier.com/authors-update/story/publishing-tips/how-to-overcome-common-obstacles-to-publishing-in-english

Veen, W., & Vrakking, B. (2006). *Homo zappiens, growing up in a digital age*. Network Continuum Education.

Ventola, E. (1992). Writing scientific English: Overcoming intercultural problems. *International Journal of Applied Linguistics*, 2(2), 191–220. doi:10.1111/j.1473-4192.1992.tb00033.x

Ventola, E. (1994). Abstracts as an object of linguistic study. In S. Čmejrková, F. Daneš, & E. Havlová (Eds.), *Writing vs. Speaking: Language, Text, Discourse, Communication* (pp. 333–352). Gunter Narr.

Vieru, D. (2015). Towards a multi-dimensional model of digital competence in small-and medium-sized enterprises. In *Encyclopedia of Information Science and Technology* (3rd ed., pp. 6715–6725). IGI Global. doi:10.4018/978-1-4666-5888-2.ch660

Vikipedi. (2020). https://tr.wikipedia.org/wiki/2019-20_koronavir%C3%BCs_pandemisi

Virtanen, M. A., Kaariainen, M., Liikanen, E., & Haavisto, E. (2017). The comparison of students' satisfaction between ubiquitous and web-based learning environments. *Education and Information Technologies*, *22*(5), 2565–2581. doi:10.100710639-016-9561-2

Vuorikari, R., Punie, Y., Carretero Gomez, S., & van den Brande, L. (2016). DigComp 2.0: The Digital Competence Framework for Citizens (Update Phase 1: The Conceptual Reference Model). Luxembourg: Publication Office of the European Union.

Wagner, T. (2008). *The global achievement gap: Why even our best schools don't teach the new survival skills our children need-and what we can do about it*. Basic Books.

Wain, K. (1993). Lifelong education: Illiberal and repressive? *Educational Philosophy and Theory*, *25*(1), 58–70. doi:10.1111/j.1469-5812.1993.tb00186.x

Wain, K. (2001). Lifelong learning: Small adjustment or paradigm shift? In *International handbook of lifelong learning* (pp. 183–198). Springer. doi:10.1007/978-94-010-0916-4_12

Waloszek, G. (2012). *SAP User Experience Community*. Introduction to Design Thinking: https://experience.sap.com/skillup/introduction-to-design-thinking/

Walter, J. B., & Burgoon, J. K. (1992). Relational Communication in Computer-Mediated Interaction. *Human Communication Research*, *19*(1), 50–88. doi:10.1111/j.1468-2958.1992.tb00295.x

Wang, K. (2012). *Digital awareness as an aspect of digital literacy*. http://as.exeter.ac.uk/cascade/digital-curriculum/case-studies/digitalawarenessasanaspectofdigitalliteracy/

Wang, Y., & Baker, R. (2015). Content or platform: Why do students complete MOOCs? *Journal of Online Learning and Teaching*, *11*(1), 17.

Wee, R. (2009). Sources of errors: An interplay of interlingual influence and intralingual factors. *European Journal of Soil Science*, *11*(2), 349–359.

White, G. K. (2013). *Digital fluency: Skills necessary for learning in the digital age*. Australian Council for Educational Research.

White, J. P. (1982). *The aims of education re-stated*. Routledge & Kegan Paul.

Williams, S. R., & Wilson, R. L. (1999). Group support systems, power & influence in an organisation: A field study. In K. E. Kendall (Ed.), *Emerging information technologies: Improving decisions, cooperation & infrastructure*. Sage. doi:10.4135/9781483345505.n8

Wilson, B., Orr, J., Hansen, D., Khoja, S., Bryant, R., & Iyer, G. (2013). MOOCs and their impact on CS education. *Journal of Comp. Sci. in Colleges*, *29*(1), 110–112.

World Bank. (2003). *Lifelong Learning in the Global Knowledge Economy: Challenges for Developing Countries*. Retrieved from https://documents.worldbank.org/curated/en/528131468749957131/Lifelong-learning-in-the-global-knowledge-economy-challenges-for-developing-countries

Wynne, B. (2008). Public participation in science and technology: Performing and Obscuring a Political-Conceptual Category Mistake. *East Asian Science, Technology and Society*, *1*(1), 99–110. doi:10.100712280-007-9004-7

Xu, Y., Park, H., & Baek, Y. (2011). A new approach toward digital storytelling: An activity focused on writing self-efficacy in a virtual learning environment. *Educational Technology & Society, 14*(4), 181-191. Retrieved from. http://www.ifets.info/journals/14_4/ets_14_4.pdf

Yamin, S., & Masek, A. (2011). The Effect of Problem Based Learning on Critical Thinking Ability: A Theoretical and Empirical Review. *International Review of Social Sciences and Humanities, 2*(1), 215–221. Retrieved November 01, 2020 form www.irssh.com

Yang, K. C. (2020). *What Can College Teachers Learn From Students' Experiential Narratives in Hybrid Courses? A Text Mining Method of Longitudinal Data. In Theoretical and Practical Approaches to Innovation in Higher Education.* IGI Global. doi:10.4018/978-1-7998-1662-1.ch006

Yang, X., Guo, X., & Yu, S. (2016). Student-generated content in college teaching: Content quality, behavioural pattern and learning performance. *Journal of Computer Assisted Learning*, *32*(1), 1–15. doi:10.1111/jcal.12111

Yıldırım, A. and Şimşek, H. (2005). *Sosyal bilimlerde nitel araştırma yöntemleri*. Ankara: Seçkin yayınları.

Yıldırım, A., & Şimşek, H. (2008). Sosyal bilimlerde nitel araştırma (6. Baskı). Ankara: Seçkin Yayınları.

Yıldırım, A., & Şimşek, H. (2006). *Sosyal bilimlerde nitel arastirma yöntemleri*. Seçkin Yayıncılık.

Yılmaz, E. (2015). *Öğretmenlerin dijital veri güvenliği farkındalığı* [Yayınlanmamış Doktora Tezi]. Anadolu Üniversitesi, Eğitim Bilimleri Enstitüsü.

Yin, R. K. (2009). *Case study research: design and methods. essential guide to qualitative methods in organizational research*. Academic Press.

Yin, R. K. (1984). *Case Study Research: Design and Methods*. Sage Publications.

Yin, R. K. (2002). *Case study research: Design and methods* (3rd ed.). Sage.

Yuan, L., & Powell, S. (2013). MOOCs and open education: Implications for higher education. Bolton, UK: JISC, Centre for Educational Technology & Interoperability Standards.

Yüksel, C. (2002). Some Thoughts on the Education Scenario - A Systems Approach. *Journal of Selcuk Communcation*, *2*, 3.

Zhang, D., Zhou, L., Briggs, R. O., & Nunamaker, J. F. Jr. (2006). Instructional video in e-learning: Assessing the impact of interactive video on learning effectiveness. *Information & Management*, *43*(1), 15–27. doi:10.1016/j.im.2005.01.004

Zu, Z. Y., Jiang, M. D., Xu, P. P., Chen, W., Ni, Q. Q., Lu, G. M., & Zhang, L. J. (2020). Coronavirus disease 2019 (COVID-19): A perspective from China. *Radiology*, *296*(2), 200490. doi:10.1148/radiol.2020200490 PMID:32083985

About the Contributors

Gulsun Kurubacak-Meric is a professor in Distance Education at the College of Open Education at Anadolu University. Dr. Meric undertook graduate studies at Anadolu University, Turkey (MA. Educational Technology) and the University of Cincinnati, USA (Ed.D. Curriculum & Instruction), and also has worked a post-doctoral fellow at the College of Education at New Mexico State University, USA (2001-2002). Dr. Meric earned her B.S. degree in Computer Engineering from the College of Informatics Technologies and Engineering of Hoca Ahmet Yesevi International Turk-Kazakhstani University in the year 2012-2013. Dr. Meric has over thirty-five years of experience in focusing on the egalitarian and ecological aspects of open and distance learning; finding new answers, viewpoints and explanations to online communication problems through critical pedagogy; and improving learner critical and creative thinking skills through project-based online learning, universal design principles and new communication technologies (ubiquities technologies, mobile technologies, virtual reality, augmented reality, mixed reality, etc.). Dr. Meric had been the Group Coordinator of R&D and International Relations at the College of Open Education of Anadolu University between 2014-2018. She has been the Quality Group Coordinator at the same College since January 2020. Dr. Gulsun Meric is the founder and editor-in-chief of AUAd (Açıköğretim Araştırmaları ve Uygulamaları Dergisi - Journal of Open Education Research and Applications - http://auad.anadolu.edu.tr) as well as associated editor of TOJDE (The Turkish Online Journal of Distance Education - http://tojde.anadolu.edu.tr). Dr. Meric is also the founding member of the Open and Distance Education Programs Accreditation Agency (AUDAK), and the head of the board of management member at the same accreditation agency. She continues to provide pedagogical support for distance learning.

Serap Sisman-Ugur graduated from an undergraduate program in "Computer Education and Instructional Technology" at Anadolu University and the master degree of the same program. Since 2002, she has been working as a lecturer at the Distance Education Department of Open Education Faculty. She works research and development activities & projects in fields such as e-learning content types, digital storytelling, animation, game-based learning, gamification, instructional design, cross-cultural aspects, artificial intelligence, individual differences and human-computer interaction. She interested in technological singularity and trans-humanism. She has been studying doctorate in Distance Learning. She was Social Media Coordinator of Anadolu University Open Education System, now she has been Social Media Coordinator and Vice of Corporate Communications Coordinator of Anadolu University.

* * *

Idris Adamu is a Lecturer at Aminu Saleh College of Education in Nigeria. He proceeds to University Technology Malaysia UTM and obtained his Master Degree in Technical and Vocational Education (Computer option) and Currently a Ph.D. candidate in Computer Education and Instructional Technology at Near East University in North Cyprus. He publishes several journals article in both Computer and Technology Education.is a Lecturer at Aminu Saleh College of Education in Nigeria. He proceeds to University Technology Malaysia UTM and obtained his Master Degree in Technical and Vocational Education (Computer option) and Currently a Ph.D. candidate in Computer Education and Instructional Technology at Near East University in North Cyprus. He publishes several journals article in both Computer and Technology Education.

Murat Aytaş was born in Istanbul in 1981. After completing his primary, secondary and high school education in the same city, he graduated from Selcuk University, Faculty of Communication, Department of Radio, Television and Cinema in 2005. In 2006, she started her graduate education at the Institute of Social Sciences, Radio-Television Department of the same university and graduated with a master's thesis entitled "The End of Space and the Transformation of the Object on Communication Technologies." In 2016, she graduated from Ege University Institute of Social Sciences, Radio-Television and Cinema Department. He graduated with his PhD thesis entitled "Visual Narration and Aesthetic Structure of Ottoman Period Series" broadcasted on Turkish Television Channels. Aytaş, who has been involved in many projects in the field of short films and documentaries since his university years, has several documentaries and short films awarded at various international and national festivals, photo exhibitions, various articles and essays on communication / cinema. A book study entitled "Images in the Purpose: Identity in the Cinema, Identity in the Cinema, and Ontological Dilemmas", published by Aytaş, who is currently working as a lecturer in the Radio-Television and Cinema Department of the Faculty of Communication at Selçuk University, has been published in 2019.

Juliana Rocha Franco is a Ph.D. in Communications and Semiotics (Pontifical Catholic University of São Paulo/ Brazil), MA in Communication and Media Studies (Federal University of Minas Gerais - Brazil) and Bachelor in History (UFMG). She is a Professor at State University of Minas Gerais (Brazil) in the Design School's graduate and undergraduate programs, and researcher at the Center for Research in Design and Ergonomics of UEMG/Brazil. Juliana has published papers and book chapters on the topics of Design, Semiotics, and Communication.

Mahmoud Hawamdeh is a Ph.D. candidate in Computer Education and Instructional Technology at Near East University in North Cyprus. He served as Director of Continuing Education and Community Service at Al-Quds Open University from January 2013 to October 2019. He was responsible for the overall direction, development, coordination, and supervision of the Center In addition to managing the proposal writing and the implementation of several projects. As a member of the team, Mr. Hawamdeh was involved in producing the six policy papers on the most successful use of ICT in student-centered learning in the Palestinian education system as a result of The Action Research conducted by the Open University of Cyprus and Al-Quds Open University consortium in 2014-15. Mr. Hawamdeh published several scientific articles and book chapters. He is the editorial board member of the Journal of International Social Studies Education (TISSE) http://tissejournal.net/, He also a SCIENCE committee member of the International Symposium of Education and Values (ISOEVA) www.isoeva.com. Moreover, he is co-founder of the Palestinian (Experts for Education) Special Interest Group-EESIG http://eesig.org/. In His professional career, Mr. Hawamdeh delivered several papers and presentations at national and international conferences. He participated as an International speaker in the UALL ANNUAL CONFERENCE 2016, held at University of Oxford (https://www.uall.ac.uk/2016-annual-conference), Guest Speaker in the International Symposium of Education and Values (ISOEVA) (http://www.isoeva.com), held by International Journal of Eurasia Social Sciences on 05-08 October 2017 in Turkey, and participated as Keynote Speaker in the 7th International Social Studies Education Symposium (http://usbes2018.ahievran.edu.tr/), which was held at Kırşehir Ahi Evran University on 11-13 October 2018.

Başak Kalkan has Bachelor's degree of Public Administration and Political Science from Istanbul University in 2002. She received her PhD from Journalism at Istanbul University in 2012. The major field of her studies is women studies, cross cultural studies, communication sociology, design of video content in distance education tools. She was a scripwriter at Department of Distance Education and Educational TV in Anadolu University Faculty of Open Education between 2010-2019. She designed and wrote many educational TV programmes. She studied on scenario design and visual content design several The Scientific And Technological Research Council of Turkey (TUBITAK) projects and Scientific Research Projects, abbreviated as BAP. She has studied to prepare in several EU projects.

İsmail Karakuş completed his undergraduate education in 2012 at Mersin University Department of Turkish Education. In 2017, he completed his master's degree in Curriculum and Instruction Department Department of Kahramanmaraş Sütçü Imam University. Then, he continued his PhD at Mersin University Curriculum and Instruction Department. He worked as a Turkish teacher at the Ministry of Education between the years 2012 and 2017. He started to work as a lecturer at Mersin University in 2018. Currently, he is working for Turkish Application and Research Center of Mersin University.

Figen Kiliç completed her undergraduate education in 1994 at Anadolu University Curriculum and Instruction Department. She worked as a research assistant at Anadolu University Faculty of Education between the years 1994-1996 and completed her master's degree at Department of Educational Sciences. She completed her PhD at the Curriculum and Instruction Department of Çukurova University in 2007. She has been working at Mersin University Department of Educational Sciences since 2007. In 2019, Figen KILIÇ became an associate professor. The author has national and international articles and books in the field of program development in education.

Hakan Kilinc is currently working as a research assistant doctor of Open and Distance Education at Open Education Faculty, Anadolu University. He undertook undergraduate studies in the feld of Computer Education and Instructional Technologies (CEIT) between the years of 2009 and 2013 at Anadolu University. He received his master degree in the feld of Distance Education in 2016, May. In addition to this, he received his Ph.D. Degree in the feld of Distance Education in 2020, April. Hakan Kılınç continues to work in the feld of distance education technology, personalized learning environments, information and communication technologies, synchronous, asynchronous, and interactive communications, learner-generated content and online group discussions.

Buket Kip Kayabaş is an assistant professor in the Department of Distance Education at Anadolu University, Open Education Faculty where she has been a faculty member since 2001. She received her bachelor degree at Anadolu University Computer Education and Instructional Technology Department in 2004. Then in 2007 she received her master degree and in 2017 Ph.D. degree in Distance Education. She has been involved in many e-learning projects in different roles such as researcher, content developer and manager. Her research interests lie in the area of openness in education, lifelong learning, open and distance learning, open educational resources and massive open online courses.

Serhat Koca completed his undergraduate education in Selcuk University, Faculty of Communication, Department of Radio, Television and Cinema, and his master's degree in the Department of Radio, Television and Cinema of the Social Sciences Institute of the same university. He graduated doctorate education at Anadolu University, Institute of Social Sciences, Department of Cinema and TV. He is still continuing studies in the field of cinema and communication sciences. Koca published his book titled "Style in New Turkish Cinema in 2018". He is currently working at Anadolu University, Faculty of Communication Sciences, Department of Cinema and Television.

Izabela Pinho is a Bachelor of Arts in Graphic Design and researcher at the Center for Research in Design and Ergonomics at State University of Minas Gerais (Brazil).

T. Volkan Yuzer, Ph.D., is professor in Distance Education at College of Open Education, Anadolu University. He undertook graduate studies at Anadolu University, Turkey. His research interests are new communication technologies, synchronous, asynchronous and interactive communications and transformative learning milieus in distance and online education. He has over twenty years' experience in exploring additional distance learning media and providing communication and technological support for distance learning programs as well as develop online learning courses. Dr. Yuzer has participated in projects related to distance learning, online synchronous learning, and virtual classroom. Dr. Yuzer is the Head of Distance Education Department of College of Open Education, Anadolu University. He has been teaching courses in distance learning, communication, and information technologies.

Index

21st century 38-40, 43, 45-47, 50, 53, 114, 116, 119, 122, 124-125, 135-136, 146, 148-149, 159, 221, 229

A

Aesthetic and Instructor Dimensions 95, 99, 101, 104, 107
Ambiguity 203, 207-208, 212, 219-220
Animation 99, 224, 240-243, 245, 247-249

C

Calvino 223, 227, 236
Cognitive Learning Theory 226
Common Errors 200, 213
communication 41, 44, 47-48, 50, 53, 57-58, 60, 62-64, 67, 69-70, 76, 78-79, 82-83, 90-91, 95-98, 100, 109, 114, 116-117, 119-121, 124, 131, 134, 147, 151, 156-157, 160-166, 170-175, 179-192, 195-196, 199, 211, 220, 222-224, 227, 229-233, 235-236, 240-241, 249
Content generating learners 133-136, 139-142, 144-146, 149
coronavirus 76, 79-81, 83, 87-88, 91, 125, 241
Covid-19 1, 11, 14, 33, 78, 83, 86, 125, 241
Creative Thinking 222

D

Deep learning 39-40, 45-47, 50-51
Design Thinking 14, 16-21, 26, 29, 34
Digital Awareness 113, 116-117, 120-123, 125, 131
Digital Competence 113-114, 117-118, 120-121, 125, 131, 160
Digital Fluency 113, 118-125, 132
digital learning environments 134, 138, 146-148, 150-152
Discourse Community 219
Distance Education 11, 14-15, 96, 98, 103, 138, 144, 146, 152, 161, 164, 166-167, 169, 173, 175, 223, 243

E

E-Certificate Programs 166-169, 172-173, 179
Emergency Remote Teaching 14-15

F

first aid 76-78, 80-83, 86, 88-91

G

Garden-Path Sentences 208, 219
Generic Errors 219
Grammatical Errors 219

H

health 4, 65, 76-82, 84, 88, 91, 116-117, 123, 125, 187-188

I

Ideo 16-17, 22
Information Design 57-60, 62-64, 66-67, 69-70

interaction 48, 80, 84, 91, 96, 100-101, 109, 122, 134, 138, 157, 163, 180-181, 183-184, 187-188, 190-191, 199, 222, 228-229, 233, 235, 249

L

Learner İdentity 223
Learning Communities 148, 156-157, 160-163, 166, 168-169, 172-175
Lexical Ambiguity 219
Lexical Errors 198, 219
Lifelong Learning 4-8, 11, 40, 53, 77-78, 96-97, 109-110, 113, 115, 117, 119-120, 122, 124-125, 132, 134, 149, 156-160, 163, 166-169, 172-174, 179

M

Massive Open Online Courses 39, 88, 163, 179
Mayer 196, 225-226, 242-243, 245-249
MOOC 88, 162, 174
multimedia learning 224, 241-243, 248-249

O

Online Learning 14-15, 21, 33, 98, 135, 162
Open and Distance Learning 152, 157, 161, 164, 172, 179

P

PBL 38-45, 48, 50-53
Post-Truth Era 1, 11
prosumer 135, 143

R

Referential Ambiguity 208, 212, 219
Research Funds 180-181, 186, 191
Research Writing 196, 199
Risk Intelligence 8-10

S

Scenario Process 221, 233
Science Communication 58, 70, 180-191, 220, 222-224, 227-228, 230-232, 235-236, 240
Scientific Communication 57-60, 63-64, 70, 76, 195-196, 199, 231-233, 249
Scientific Writing 195-197, 199-200, 203, 207, 211-213
Social Media 2-3, 76-77, 79-80, 82-84, 90-92, 181, 184, 189-190
Social Responsibility 76, 78, 84, 88
Stanford University 23, 25
Storification 232
storytelling 29, 70, 232, 240, 243, 245
Student Satisfaction 95, 97, 101
Syntactic Ambiguity 219

T

Template Analysis 196, 199-200

U

UNESCO 14-15, 77, 159

V

Vagueness 219
Visual Communication 57, 59, 241, 249

www.ingramcontent.com/pod-product-compliance
Lightning Source LLC
LaVergne TN
LVHW081332021026
849161LV00008BA/1953